Defoe Abbott Marx Hardy Melville Montaigne Machiavelli Haggard Chesterton Cooper Emerson Joyce Austen Hugo Eliot Grimm
Stoker Carroll Christie Maupassant Byron Engels Schiller Molière
Wilde Garnett Einstein Fitzgerald Hawthorne Smith Kafka Hall
Goethe Cotton Dostoyevsky Willis
Baum Henry Kipling Doyle
Leslie Dumas Flaubert Turgenev Nietzsche Balzac Crane
Stockton Vatsyayana Verne Vinci
Burroughs Whitman Gogol Busch
Curtis Tocqueville
Homer Widger Tolstoy Twain Scott Plato Harte
Darwin Thoreau
Potter Freud Zola Lawrence Dickens Burton Hesse
Kant Jowett Stevenson
Andersen
London Descartes Cervantes Voltaire
Poe Aristotle Wells Cooke
Hale James Hastings Irving
Bunner Shakespeare
Richter Dante Chambers da Shaw Benedict Alcott
Doré Swift Chekhov Wodehouse Pushkin Newton

tredition

tredition was established in 2006 by Sandra Latusseck and Soenke Schulz. Based in Hamburg, Germany, tredition offers publishing solutions to authors and publishing houses, combined with world-wide distribution of printed and digital book content. tredition is uniquely positioned to enable authors and publishing houses to create books on their own terms and without conventional manu-facturing risks.

For more information please visit: www.tredition.com

TREDITION CLASSICS

This book is part of the TREDITION CLASSICS series. The creators of this series are united by passion for literature and driven by the intention of making all public domain books available in printed format again - worldwide. Most TREDITION CLASSICS titles have been out of print and off the bookstore shelves for decades. At tredition we believe that a great book never goes out of style and that its value is eternal. Several mostly non-profit literature projects provide content to tredition. To support their good work, tredition donates a portion of the proceeds from each sold copy. As a reader of a TREDITION CLASSICS book, you support our mission to save many of the amazing works of world literature from oblivion. See all available books at www.tredition.com.

 Project Gutenberg

The content for this book has been graciously provided by Project Gutenberg. Project Gutenberg is a non-profit organization founded by Michael Hart in 1971 at the University of Illinois. The mission of Project Gutenberg is simple: To encourage the creation and distribution of eBooks. Project Gutenberg is the first and largest collection of public domain eBooks.

A Treatise on Domestic Economy For the Use of Young Ladies at Home and at School

Catharine Esther Beecher

Imprint

This book is part of TREDITION CLASSICS

Author: Catharine Esther Beecher
Cover design: Buchgut, Berlin – Germany

Publisher: tredition GmbH, Hamburg - Germany
ISBN: 978-3-8472-2765-6

www.tredition.com
www.tredition.de

TO

AMERICAN MOTHERS,

whose intelligence and virtues have inspired admiration and respect, whose experience has furnished many valuable suggestions, in this work, whose approbation will be highly valued, and whose influence, in promoting the object aimed at, is respectfully solicited, this work is dedicated, by their friend and countrywoman,

THE AUTHOR.

PREFACE
TO THE THIRD EDITION.

The author of this work was led to attempt it, by discovering, in her extensive travels, the deplorable sufferings of multitudes of young wives and mothers, from the combined influence of *poor health, poor domestics, and a defective domestic education*. The number of young women whose health is crushed, ere the first few years of married life are past, would seem incredible to one who has not investigated this subject, and it would be vain to attempt to depict the sorrow, discouragement, and distress experienced in most families where the wife and mother is a perpetual invalid.

The writer became early convinced that this evil results mainly from the fact, that young girls, especially in the more wealthy classes, *are not trained for their profession*. In early life, they go through a course of school training which results in great debility of constitution, while, at the same time, their physical and domestic education is almost wholly neglected. Thus they enter on their most arduous and sacred duties so inexperienced and uninformed, and with so little muscular and nervous strength, that probably there is not *one chance in ten*, that young women of the present day, will pass through the first years of married life without such prostration of health and spirits as makes life a burden to themselves, and, it is to be feared, such as seriously interrupts the confidence and happiness of married life.

The measure which, more than any other, would tend to remedy this evil, [Pg 6]would be to place *domestic economy* on an equality with the other sciences in female schools. This should be done because it *can* be properly and systematically taught (not *practically*, but as a *science*), as much so as *political economy* or *moral science*, or any other branch of study; because it embraces knowledge, which will be needed by young women at all times and in all places; because this science can never be *properly* taught until it is made a branch of *study*; and because this method will secure a dignity and importance in the estimation of young girls, which can never be accorded while they perceive their teachers and parents practically attaching more value to every other department of science than this.

When young ladies are taught the construction of their own bodies, and all the causes in domestic life which tend to weaken the constitution; when they are taught rightly to appreciate and learn the most convenient and economical modes of performing all family duties, and of employing time and money; and when they perceive the true estimate accorded to these things by teachers and friends, the grand cause of this evil will be removed. Women will be trained to secure, as of first importance, a strong and healthy constitution, and all those rules of thrift and economy that will make domestic duty easy and pleasant.

To promote this object, the writer prepared this volume as a *text-book* for female schools. It has been examined by the Massachusetts Board of Education, and been deemed worthy by them to be admitted as a part of the Massachusetts School Library.

It has also been adopted as a text-book in some of our largest and most popular female schools, both at the East and West.

The following, from the pen of Mr. George B. Emmerson, one of the most popular and successful teachers in our country, who has introduced this work as a text-book in [Pg 7]his own school, will exhibit the opinion of one who has formed his judgment from experience in the use of the work:

"It may be objected that such things cannot be taught by books. Why not? Why may not the structure of the human body, and the laws of health deduced therefrom, be as well taught as the laws of natural philosophy? Why are not the application of these laws to the management of infants and young children as important to a woman as the application of the rules of arithmetic to the extraction of the cube root? Why may not the properties of the atmosphere be explained, in reference to the proper ventilation of rooms, or exercise in the open air, as properly as to the burning of steel or sodium? Why is not the human skeleton as curious and interesting as the air-pump; and the action of the brain, as the action of a steam-engine? Why may not the healthiness of different kinds of food and drink, the proper modes of cooking, and the rules in reference to the modes and times of taking them, be discussed as properly as rules of grammar, or facts in history? Are not the principles that should regulate clothing, the rules of cleanliness, the advantages of early rising and domestic exercise, as readily communicated as the principles of mineralogy, or rules of syntax? Are not the rules of Jesus Christ, applied to refine *domestic manners* and preserve a *good temper*, as important as the abstract principles of ethics, as taught by Paley, Wayland, or Jouffroy? May not the advantages of neatness, system, and order, be as well illustrated in showing how they contribute to the happiness of a family, as by showing how they add beauty to a copy-book, or a portfolio of drawings? Would not a teacher be as well employed in teaching the rules of economy, in regard to time and expenses, or in regard to dispensing charity, as in teaching double, or single entry in bookkeeping? [Pg 8]Are not the principles that should guide in constructing a house, and in warming and ventilating it properly, as important to young girls as the principles of the Athenian Commonwealth, or the rules of Roman tactics? Is it not as important that children should be taught the dangers to the mental faculties, when over-excited on the one hand, or left unoccupied on the other, as to teach them the conflicting theories of political economy, or the speculations of metaphysicians? For ourselves, we have always found children, especially girls, peculiarly ready to listen to what they saw would prepare them for future duties. The truth, that education should be *a preparation for actual, real life*, has the greatest force with children. The constantly-recurring inquiry, 'What will be *the use* of this study?' is always satisfied by showing, that it will prepare for any duty, relation, or office which, in the natural course of things, will be likely to come.

"We think this book extremely well suited to be used as a text-book in schools for young ladies, and many chapters are well adapted for a reading book for children of both sexes."

To this the writer would add the testimony of a lady who has used this work with several classes of young girls and young ladies. She remarked that she had never known a school-book that awakened more interest, and that some young girls would learn a lesson in this when they would study nothing else. She remarked, also, that when reciting the chapter on the construction of houses, they became greatly interested in inventing plans of their own, which gave an opportunity to the teacher to point out difficulties and defects. Had this part of domestic economy been taught in schools, our land would not be so defaced with awkward, misshapen, inconvenient, and, at the same time, needlessly expensive houses, as it now is.

[Pg 9]

Although the writer was trained to the care of children, and to perform all branches of domestic duty, by some of the best of housekeepers, much in these pages is offered, not as the result of her own experience, but as what has obtained the approbation of some of the most judicious mothers and housekeepers in the nation. The articles on Physiology and Hygiene, and those on horticulture, were derived from standard works on these subjects, and are sanctioned by the highest authorities.

The American Housekeeper's Receipt Book is another work prepared by the author of the Domestic Economy, in connexion with several experienced housekeepers, and is designed for a supplement to this work. On pages 354a and 354b will be found the Preface and Analysis of that work, the two books being designed for a complete course of instructions on every department of Domestic Economy.

The copyright interest in these two works is held by a board of gentlemen appointed for the purpose, who, after paying a moderate compensation to the author for the time and labour spent in preparing these works, will employ all the remainder paid over by the publishers, to aid in educating and locating such female teachers as wish to be employed in those portions of our country, which are most destitute of schools.

The contract with the publisher provides that the publisher shall guaranty the sales and thus secure against any losses for bad debts, for which he shall receive five *per cent*. He shall charge twenty per cent. for commissions paid to retailers, and also the expenses of printing, paper, and binding, at the current market prices, and make no other charges. The net profits thus determined are then to be divided equally, the publishers taking one half, and paying the other half to the board above mentioned.

[Pg 11]

CONTENTS.

CHAPTER III.

REMEDIES FOR THE PRECEDING DIFFICULTIES.

CHAPTER IV.

ON DOMESTIC ECONOMY AS A BRANCH OF STUDY.

CHAPTER V.

ON THE CARE OF HEALTH.

Importance of a Knowledge of the Laws of Health, and of the Human System, to Females. Construction of the Human Frame. Bones; their Structure, Design, and Use. Engraving and Description. Spinal Column. Engravings of Vertebræ. Exercise of the Bones. Muscles; their Constitution, Use, and Connection with the Bones. Engraving and Description. Operation of Muscles. Nerves; their Use. Spinal Column. Engravings and Descriptions. Distortions of the Spine. Engravings and Descriptions. Blood-Vessels; their Object. Engravings and Descriptions. The Heart, and its Connection with the [Pg 13]System. Engravings and Descriptions. Organs of Digestion and Respiration. Engraving and Description. Process of Digestion. Circulation of the Blood. Process of Respiration. Necessity of Pure Air. The Skin. Process of Perspiration. Insensible Perspiration. Heat of the Body. Absorbents. Importance of frequent Ablutions and Change of Garments. Follicles of Oily Matter in the Skin. Nerves of Feeling,

CHAPTER VI.

ON HEALTHFUL FOOD.

Responsibility of a Housekeeper in Regard to Health and Food. The most fruitful Cause of Disease. Gastric Juice; how proportioned. Hunger the Natural Guide as to Quantity of Food. A Benevolent Provision; how perverted, and its Effects. A Morbid Appetite, how caused. Effects of too much Food in the Stomach. Duty of a Housekeeper in Reference to this. Proper Time for taking Food. Peristaltic Motion. Need of Rest to the Muscles of the Stomach. Time necessary between each Meal. Exceptions of hard Laborers and active Children. Exercise; its Effect on all parts of the Body. How it produces Hunger. What is to be done by those who have lost the Guidance of Hunger in regulating the Amount of Food. On Quality of Food. Difference as to Risk from bad Food, between Healthy Persons who exercise, and those of Delicate and Sedentary Habits. Stimulating Food; its Effects. Condiments needed only for Medicine, and to be avoided as Food. Difference between Animal and Vegetable Food.

CHAPTER VII.

ON HEALTHFUL DRINKS.

CHAPTER VIII.

ON CLOTHING.

CHAPTER IX.

ON CLEANLINESS.

CHAPTER X.

ON EARLY RISING.

CHAPTER XIII.

ON THE PRESERVATION OF A GOOD TEMPER IN A HOUSE-KEEPER.

CHAPTER XIV.

ON HABITS OF SYSTEM AND ORDER.

CHAPTER XV.

ON GIVING IN CHARITY.

CHAPTER XVI.

ON ECONOMY OF TIME AND EXPENSES.

Economy of Time. Value of Time. Right Apportionment of Time. Laws appointed by God for the Jews. Proportions of Property and Time the Jews were required to devote to Intellectual, Benevolent, and Religious Purposes. The Levites. The weekly Sabbath. The Sabbatical Year. Three sevenths of the Time of the Jews devoted to God's Service. Christianity removes the Restrictions laid on the Jews, but demands all our Time to be devoted to our own best Interests and the Good of our Fellow-men. Some Practical Good to be the Ultimate End of all our Pursuits. Enjoyment connected with the Performance of every Duty. Great Mistake of Mankind. A Final Account to be given of the Apportionment of our Time. Various Modes of economizing Time. System and Order. Uniting several Objects in one Employment. Employment of Odd Intervals of Time. We are bound to aid Others in economizing Time. *Economy in Expenses.* Necessity of Information on this Point. Contradictory Notions. General Principles in which all agree. Knowledge of Income and Expenses. Every One bound to do as much as she can to secure System and Order. Examples. Evils of Want of System and Forethought. Young [Pg 18]Ladies should early learn to be systematic and economical. Articles of Dress and Furniture should be in Keeping with each other, and with the Circumstances of the Family. Mistaken Economy. Education of Daughters away from Home injudicious. Nice Sewing should be done at Home. Cheap Articles not always most economical. Buying by wholesale economical only in special cases. Penurious Savings made by getting the Poor to work cheap. Relative Obligations of the Poor and the Rich in Regard to Economy. Economy of Providence in the Unequal Distribution of Property. Carelessness of Expense not a Mark of Gentility. Beating down Prices improper in Wealthy People. Inconsistency in American would-be Fashionables,

19]Preventing, better than finding Fault. Faults should be pointed out in a Kind Manner. Some Employers think it their Office and Duty to find Fault. Domestics should be regarded with Sympathy and Forbearance,

CHAPTER XIX.

ON THE CARE OF INFANTS.

Necessity of a Knowledge of this Subject, to every Young Lady. Examples. Extracts from Doctors Combe, Bell, and Eberle. Half the Deaths of Infants owing to Mismanagement, and Errors in Diet. Errors of Parents and Nurses. Error of administering Medicines to Children, unnecessarily. Need of Fresh Air, Attention to Food, Cleanliness, Dress, and Bathing. Cholera Infantum not cured by Nostrums. Formation of Good Habits in Children,

CHAPTER XX.

ON THE MANAGEMENT OF YOUNG CHILDREN.

Physical Education of Children. Remark of Dr. Clark, and Opinion of other Medical Men. Many Popular Notions relating to Animal Food for Children, erroneous. The Formation of the Human Teeth and Stomach does not indicate that Man was designed to live on Flesh. Opinions of Linnæus and Cuvier. Stimulus of Animal Food not necessary to Full Developement of the Physical and Intellectual Powers. Examples. Of Laplanders, Kamtschatkadales, Scotch Highlanders, Siberian Exiles, Africans, Arabs. Popular Notion that Animal Food is more Nourishing than Vegetable. Different Opinions on this Subject. Experiments. Opinions of Dr. Combe and others. Examples of Men who lived to a great Age. Dr. Franklin's Testimony. Sir Isaac Newton and others. Albany Orphan Asylum. Deleterious Practice of allowing Children to eat at short Intervals. Intellectual Training. Schoolrooms. Moral Character. Submission, Self-denial, and Benevolence, the three most important Habits to be formed in Early Life. Extremes to be guarded against. Medium Course. Adults sometimes forget the Value which Children set on Trifles. Example. Impossible to govern Children, properly, without appreciating the

CHAPTER XXI.

ON THE CARE OF THE SICK.

CHAPTER XXII.

ON ACCIDENTS AND ANTIDOTES.

Head. Burns. Drowning. Poisons:—Corrosive Sublimate; Arsenic, or Cobalt; Opium; Acids; Alkalies. Stupefaction from Fumes of Charcoal, or from entering a Well, Limekiln, or Coalmine. Hemorrhage of the Lungs, Stomach, or Throat. Bleeding of the Nose. Dangers from Lightning,

CHAPTER XXIII.

ON DOMESTIC AMUSEMENTS AND SOCIAL DUTIES.

Indefiniteness of Opinion on this Subject. Every Person needs some Recreation. General Rules. How much Time to be given. What Amusements proper. Those should always be avoided, which cause Pain, or injure the Health, or endanger Life, or interfere with important Duties, or are pernicious in their Tendency. Horse-racing, Circus-riding, Theatres, and Gambling. Dancing, as now conducted, does not conduce to Health of Body or Mind, but the contrary. Dancing in the Open Air beneficial. Social Benefits of Dancing considered. Ease and Grace of Manners better secured by a System of Calisthenics. The Writer's Experience. Balls going out of Fashion, among the more refined Circles. Novel-reading. Necessity for Discrimination. Young Persons should be guarded from Novels. Proper Amusements for Young Persons. Cultivation of Flowers and Fruits. Benefits of the Practice. Music. Children enjoy it. Collections of [Pg 21]Shells, Plants, Minerals, &c. Children's Games and Sports. Parents should join in them. Mechanical Skill of Children to be encouraged. Other Enjoyments. Social Enjoyments not always considered in the List of Duties. Main Object of Life to form Character. Family Friendship should be preserved. Plan adopted by Families of the Writer's Acquaintance. Kindness to Strangers. Hospitality. Change of Character of Communities in Relation to Hospitality. Hospitality should be prompt. Strangers should be made to feel at their Ease,

CHAPTER XXIV.

ON THE CONSTRUCTION OF HOUSES.

Importance to Family Comfort of well-constructed Houses. Rules for constructing them. Economy of Labor. Large Houses. Arrangement of Rooms. Wells and Cisterns. Economy of Money. Shape and Arrangement of Houses. Porticoes, Piazzas, and other Ornaments. Simplicity to be preferred. Fireplaces. Economy of Health. Outdoor Conveniences. Doors and Windows. Ventilation. Economy of Comfort. Domestics. Spare Chambers. Good Taste. Proportions. Color and Ornaments. *Plans of Houses and Domestic Conveniences*. Receipts for Whitewash,

CHAPTER XXV.

ON FIRES AND LIGHTS.

Wood Fires. Construction of Fireplaces. Firesets. Building a Fire. Wood. Cautions. Stoves and Grates. Cautions. Stovepipes. Anthracite Coal. Bituminous Coal. Proper Grates. Coal Stoves. *On Lights*. Lamps. Oil. Candles. Lard. Pearlash and Water for cleansing Lamps. Care of Lamps. Difficulty. Articles needed in trimming Lamps. Astral Lamps. Wicks. Dipping Wicks in Vinegar. Shades. Weak Eyes. Entry Lamps. Night Lamps. Tapers. Wax Tapers for Use in Sealing Letters. To make Candles. Moulds. Dipped Candles. Rush Lights,

CHAPTER XXVI.

ON WASHING.

All needful Accommodations should be provided. Plenty of Water, easily accessible, necessary. Articles to be provided for Washing. Substitutes for Soft Water. Common Mode of Washing. Assorting Clothes. To Wash Bedding. Feathers. Calicoes. Bran-water. Potato-water. Soda Washing. Soda Soap. Mode of Soda Washing. Cautions in Regard to Colored Clothes, and Flannels. To Wash Brown Linen, Muslins, Nankeen, Woollen Table-Covers and Shawls, Woollen Yarn, Worsted and Woollen Hose. To Cleanse Gentlemen's Broadcloths. To make Ley, Soft Soap, Hard Soap, White Soap, Starch, and other Articles used in Washing,

LIST OF ENGRAVINGS.

[Pg 25]

DOMESTIC ECONOMY.

CHAPTER I.
THE PECULIAR RESPONSIBILITIES OF AMERICAN WOMEN.

There are some reasons, why American women should feel an interest in the support of the democratic institutions of their Country, which it is important that they should consider. The great maxim, which is the basis of all our civil and political institutions, is, that "all men are created equal," and that they are equally entitled to "life, liberty, and the pursuit of happiness."

But it can readily be seen, that this is only another mode of expressing the fundamental principle which the Great Ruler of the Universe has established, as the law of His eternal government. "Thou shalt love thy neighbor as thyself;" and "Whatsoever ye would that men should do to you, do ye even so to them," are the Scripture forms, by which the Supreme Lawgiver requires that each individual of our race shall regard the happiness of others, as of the same value as his own; and which forbid any institution, in private or civil life, which secures advantages to one class, by sacrificing the interests of another.

The principles of democracy, then, are identical with the principles of Christianity.

But, in order that each individual may pursue and secure the highest degree of happiness within his reach, unimpeded by the selfish interests of others, a system of laws must be established, which sustain certain relations and dependencies in social and civil life. What these relations and their attending obligations shall be, are to [Pg 26]be determined, not with reference to the wishes and interests of a few, but solely with reference to the general good of all; so that each individual shall have his own interest, as well as the public benefit, secured by them.

For this purpose, it is needful that certain relations be sustained, which involve the duties of subordination. There must be the magistrate and the subject, one of whom is the superior, and the other the inferior. There must be the relations of husband and wife, parent and child, teacher and pupil, employer and employed, each involving the relative duties of subordination. The superior, in certain particulars, is to direct, and the inferior is to yield obedience. Society could never go forward, harmoniously, nor could any craft or profession be successfully pursued, unless these superior and subordinate relations be instituted and sustained.

But who shall take the higher, and who the subordinate, stations in social and civil life? This matter, in the case of parents and children, is decided by the Creator. He has given children to the control of parents, as their superiors, and to them they remain subordinate, to a certain age, or so long as they are members of their household. And parents can delegate such a portion of their authority to teachers and employers, as the interests of their children require.

In most other cases, in a truly democratic state, each individual is allowed to choose for himself, who shall take the position of his superior. No woman is forced to obey any husband but the one she chooses for herself; nor is she obliged to take a husband, if she prefers to remain single. So every domestic, and every artisan or laborer, after passing from parental control, can choose the employer to whom he is to accord obedience, or, if he prefers to relinquish certain advantages, he can remain without taking a subordinate place to any employer.

Each subject, also, has equal power with every other, to decide who shall be his superior as a ruler. The weakest, the poorest, the most illiterate, has the same [Pg 27]opportunity to determine this question, as the richest, the most learned, and the most exalted.

And the various privileges that wealth secures, are equally open to all classes. Every man may aim at riches, unimpeded by any law or institution which secures peculiar privileges to a favored class, at the expense of another. Every law, and every institution, is tested by examining whether it secures equal advantages to all; and, if the people become convinced that any regulation sacrifices the good of the majority to the interests of the smaller number, they have power to abolish it.

The institutions of monarchical and aristocratic nations are based on precisely opposite principles. They secure, to certain small and favored classes, advantages, which can be maintained, only by sacrificing the interests of the great mass of the people. Thus, the throne and aristocracy of England are supported by laws and customs, which burden the lower classes with taxes, so enormous, as to deprive them of all the luxuries, and of most of the comforts, of life. Poor dwellings, scanty food, unhealthy employments, excessive labor, and entire destitution of the means and time for education, are appointed for the lower classes, that a few may live in palaces, and riot in every indulgence.

The tendencies of democratic institutions, in reference to the rights and interests of the female sex, have been fully developed in the United States; and it is in this aspect, that the subject is one of peculiar interest to American women. In this Country, it is established, both by opinion and by practice, that woman has an equal interest in all social and civil concerns; and that no domestic, civil, or political, institution, is right, which sacrifices her interest to promote that of the other sex. But in order to secure her the more firmly in all these privileges, it is decided, that, in the domestic relation, she take a subordinate station, and that, in civil and political concerns, her interests be intrusted to the other sex, without her taking any part in voting, or in making and administering laws. The result of this order of [Pg 28]things has been fairly tested, and is thus portrayed by M. De Tocqueville, a writer, who, for intelligence, fidelity, and ability, ranks second to none.

"There are people in Europe, who, confounding together the different characteristics of the sexes, would make of man and woman, beings not only equal, but alike. They would give to both the same functions, impose on both the same duties, and grant to both the same rights. They would mix them in all things,—their business, their occupations, their pleasures. It may readily be conceived, that, by *thus* attempting to make one sex equal to the other, both are degraded; and, from so preposterous a medley of the works of Nature, nothing could ever result, but weak men and disorderly women.

"It is not thus that the Americans understand the species of democratic equality, which may be established between the sexes. They admit, that, as Nature has appointed such wide differences between the physical and moral constitutions of man and woman, her manifest design was, to give a distinct employment to their various faculties; and they hold, that improvement does not consist in making beings so dissimilar do pretty nearly the same things, but in getting each of them to fulfil their respective tasks, in the best possible manner. The Americans have applied to the sexes the great principle of political economy, which governs the manufactories of our age, by carefully dividing the duties of man from those of woman, in order that the great work of society may be the better carried on.

"In no country has such constant care been taken, as in America, to trace two clearly distinct lines of action for the two sexes, and to make them keep pace one with the other, but in two pathways which are always different. American women never manage the outward concerns of the family, or conduct a business, or take a part in political life; nor are they, on the other hand, ever compelled to perform the rough labor of the fields, or to make any of those laborious exertions, [Pg 29]which demand the exertion of physical strength. No families are so poor, as to form an exception to this rule.

"If, on the one hand, an American woman cannot escape from the quiet circle of domestic employments, on the other hand, she is never forced to go beyond it. Hence it is, that the women of America, who often exhibit a masculine strength of understanding, and a manly energy, generally preserve great delicacy of personal appearance, and always retain the manners of women, although they sometimes show that they have the hearts and minds of men.

"Nor have the Americans ever supposed, that one consequence of democratic principles, is, the subversion of marital power, or the confusion of the natural authorities in families. They hold, that every association must have a head, in order to accomplish its object; and that the natural head of the conjugal association is man. They do not, therefore, deny him the right of directing his partner; and they maintain, that, in the smaller association of husband and wife, as well as in the great social community, the object of democracy is, to regulate and legalize the powers which are necessary, not to subvert all power.

"This opinion is not peculiar to one sex, and contested by the other. I never observed, that the women of America considered conjugal authority as a fortunate usurpation of their rights, nor that they thought themselves degraded by submitting to it. It appears to me, on the contrary, that they attach a sort of pride to the voluntary surrender of their own will, and make it their boast to bend themselves to the yoke, not to shake it off. Such, at least, is the feeling expressed by the most virtuous of their sex; the others are silent; and in the United States it is not the practice for a guilty wife to clamor for the rights of woman, while she is trampling on her holiest duties."

"Although the travellers, who have visited North America, differ on a great number of points, they agree in remarking, that morals are far more strict, there, than [Pg 30]elsewhere.[A] It is evident that, on this point, the Americans are very superior to their progenitors, the English." "In England, as in all other Countries of Europe, public malice is constantly attacking the frailties of women. Philosophers and statesmen are heard to deplore, that morals are not sufficiently strict; and the literary productions of the Country constantly lead one to suppose so. In America, all books, novels not excepted, suppose women to be chaste; and no one thinks of relating affairs of gallantry."

"It has often been remarked, that, in Europe, a certain degree of contempt lurks, even in the flattery which men lavish upon women. Although a European frequently affects to be the slave of woman, it may be seen, that he never sincerely thinks her his equal. In the United States, men seldom compliment women, but they daily show how much they esteem them. They constantly display an entire confidence in the understanding of a wife, and a profound respect for her freedom."

[Pg 31]

They have decided that her mind is just as fitted as that of a man to discover the plain truth, and her heart as firm to embrace it, and they have never sought to place her virtue, any more than his, under the shelter of prejudice, ignorance, and fear.

"It would seem, that in Europe, where man so easily submits to the despotic sway of woman, they are nevertheless curtailed of some of the greatest qualities of the human species, and considered as seductive, but imperfect beings, and (what may well provoke astonishment) women ultimately look upon themselves in the same light, and almost consider it as a privilege that they are entitled to show themselves futile, feeble, and timid. The women of America claim no such privileges."

"It is true, that the Americans rarely lavish upon women those eager attentions which are commonly paid [Pg 32]them in Europe. But their conduct to women always implies, that they suppose them to be virtuous and refined; and such is the respect entertained for the moral freedom of the sex, that, in the presence of a woman, the most guarded language is used, lest her ear should be offended by an expression. In America, a young unmarried woman may, alone, and without fear, undertake a long journey."

"Thus the Americans do not think that man and woman have either the duty, or the right, to perform the same offices, but they show an equal regard for both their respective parts; and, though their lot is different, they consider both of them, as beings of equal value. They do not give to the courage of woman the same form, or the same direction, as to that of man; but they never doubt her courage: and if they hold that man and his partner ought not always to exercise their intellect and understanding in the same manner, they at least believe the understanding of the one to be as sound as that of the other, and her intellect to be as clear. Thus, then, while they have allowed the social inferiority of woman to subsist, they have done all they could to raise her, morally and intellectually, to the level of man; and, in this respect, they appear to me to have excellently understood the true principle of democratic improvement.

"As for myself, I do not hesitate to avow, that, although the women of the United States are confined within the narrow circle of domestic life, and their situation is, in some respects, one of extreme dependence, I have nowhere seen women occupying a loftier position; and if I were asked, now I am drawing to the close of this work, in which I have spoken of so many important things done by the Americans, to what the singular prosperity and growing strength of that people ought mainly to be attributed, I should reply, — *to the superiority of their women.*"

This testimony of a foreigner, who has had abundant opportunities of making a comparison, is sanctioned by [Pg 33]the assent of all candid and intelligent men, who have enjoyed similar opportunities.

It appears, then, that it is in America, alone, that women are raised to an equality with the other sex; and that, both in theory and practice, their interests are regarded as of equal value. They are made subordinate in station, only where a regard to their best interests demands it, while, as if in compensation for this, by custom and courtesy, they are always treated as superiors. Universally, in this Country, through every class of society, precedence is given to woman, in all the comforts, conveniences, and courtesies, of life.

In civil and political affairs, American women take no interest or concern, except so far as they sympathize with their family and personal friends; but in all cases, in which they do feel a concern, their opinions and feelings have a consideration, equal, or even superior, to that of the other sex.

In matters pertaining to the education of their children, in the selection and support of a clergyman, in all benevolent enterprises, and in all questions relating to morals or manners, they have a superior influence. In such concerns, it would be impossible to carry a point, contrary to their judgement and feelings; while an enterprise, sustained by them, will seldom fail of success.

If those who are bewailing themselves over the fancied wrongs and injuries of women in this Nation, could only see things as they are, they would know, that, whatever remnants of a barbarous or aristocratic age may remain in our civil institutions, in reference to the interests of women, it is only because they are ignorant of them, or do not use their influence to have them rectified; for it is very certain that there is nothing reasonable, which American women would unite in asking, that would not readily be bestowed.

The preceding remarks, then, illustrate the position, that the democratic institutions of this Country are in [Pg 34]reality no other than the principles of Christianity carried into operation, and that they tend to place woman in her true position in society, as having equal rights with the other sex; and that, in fact, they have secured to American women a lofty and fortunate position, which, as yet, has been attained by the women of no other nation.

There is another topic, presented in the work of the above author, which demands the profound attention of American women.

The following is taken from that part of the Introduction to the work, illustrating the position, that, for ages, there has been a constant progress, in all civilized nations, towards the democratic equality attained in this Country.

"The various occurrences of national existence have every where turned to the advantage of democracy; all men have aided it by their exertions; those who have intentionally labored in its cause, and those who have served it unwittingly; those who have fought for it, and those who have declared themselves its opponents, have all been driven along in the same track, have all labored to one end;" "all have been blind instruments in the hands of God."

"The gradual developement of the equality of conditions, is, therefore, a Providential fact; and it possesses all the characteristics of a Divine decree: it is universal, it is durable, it constantly eludes all human interference, and all events, as well as all men, contribute to its progress."

"The whole book, which is here offered to the public, has been written under the impression of a kind of religious dread, produced in the author's mind, by the contemplation of so irresistible a revolution, which has advanced for centuries, in spite of such amazing obstacles, and which is still proceeding in the midst of the ruins it has made.

"It is not necessary that God Himself should speak, [Pg 35]in order to disclose to us the unquestionable signs of His will. We can discern them in the habitual course of Nature, and in the invariable tendency of events."

"If the men of our time were led, by attentive observation, and by sincere reflection, to acknowledge that the gradual and progressive developement of social equality is at once the past and future of their history, this solitary truth would confer the sacred character of a Divine decree upon the change. To attempt to check democracy, would be, in that case, to resist the will of God; and the nations

would then be constrained to make the best of the social lot award-
ed to them by Providence."

"It is not, then, merely to satisfy a legitimate curiosity, that I have
examined America; my wish has been to find instruction by which
we may ourselves profit." "I have not even affected to discuss
whether the social revolution, which I believe to be irresistible, is
advantageous or prejudicial to mankind. I have acknowledged this
revolution, as a fact already accomplished, or on the eve of its accomplishment; and I have selected the nation, from among those
which have undergone it, in which its developement has been the
most peaceful and the most complete, in order to discern its natural
consequences, and, if it be possible, to distinguish the means by
which it may be rendered profitable. I confess, that in America I saw
more than America; I sought the image of democracy itself, with its
inclinations, its character, its prejudices, and its passions, in order to
learn what we have to fear, or to hope, from its progress."

It thus appears, that the sublime and elevating anticipations
which have filled the mind and heart of the religious world, have
become so far developed, that philosophers and statesmen are per-
ceiving the signs, and are predicting the approach, of the same
grand consummation. There is a day advancing, "by seers predicted,
and by poets sung," when the curse of selfishness shall be removed;
when "scenes surpassing fable, [Pg 36]and yet true," shall be realized; when all nations shall rejoice and be made blessed, under
those benevolent influences, which the Messiah came to establish on
earth.

And this is the Country, which the Disposer of events designs
shall go forth as the cynosure of nations, to guide them to the light
and blessedness of that day. To us is committed the grand, the responsible privilege, of exhibiting to the world, the beneficent influences of Christianity, when carried into every social, civil, and political institution; and, though we have, as yet, made such imperfect
advances, already the light is streaming into the dark prison-house
of despotic lands, while startled kings and sages, philosophers and
statesmen, are watching us with that interest, which a career so
illustrious, and so involving their own destiny, is calculated to excite. They are studying our institutions, scrutinizing our experience,

and watching for our mistakes, that they may learn whether "a social revolution, so irresistible, be advantageous or prejudicial to mankind."

There are persons, who regard these interesting truths merely as food for national vanity; but every reflecting and Christian mind, must consider it as an occasion for solemn and anxious reflection. Are we, then, a spectacle to the world? Has the Eternal Lawgiver appointed us to work out a problem, involving the destiny of the whole earth? Are such momentous interests to be advanced or retarded, just in proportion as we are faithful to our high trust? "What manner of persons, then, ought we to be," in attempting to sustain so solemn, so glorious a responsibility?

But the part to be enacted by American women, in this great moral enterprise, is the point to which special attention should here be directed.

The success of democratic institutions, as is conceded by all, depends upon the intellectual and moral character of the mass of the people. If they are intelligent [Pg 37]and virtuous, democracy is a blessing; but if they are ignorant and wicked, it is only a curse, and as much more dreadful than any other form of civil government, as a thousand tyrants are more to be dreaded than one. It is equally conceded, that the formation of the moral and intellectual character of the young is committed mainly to the female hand. The mother forms the character of the future man; the sister bends the fibres that are hereafter to be the forest tree; the wife sways the heart, whose energies may turn for good or for evil the destinies of a nation. Let the women of a country be made virtuous and intelligent, and the men will certainly be the same. The proper education of a man decides the welfare of an individual; but educate a woman, and the interests of a whole family are secured.

If this be so, as none will deny, then to American women, more than to any others on earth, is committed the exalted privilege of extending over the world those blessed influences, which are to renovate degraded man, and "clothe all climes with beauty."

No American woman, then, has any occasion for feeling that hers is an humble or insignificant lot. The value of what an individual accomplishes, is to be estimated by the importance of the enterprise achieved, and not by the particular position of the laborer. The drops of heaven which freshen the earth, are each of equal value, whether they fall in the lowland meadow, or the princely parterre. The builders of a temple are of equal importance, whether they labor on the foundations, or toil upon the dome.

Thus, also, with those labors which are to be made effectual in the regeneration of the Earth. And it is by forming a habit of regarding the apparently insignificant efforts of each isolated laborer, in a comprehensive manner, as indispensable portions of a grand result, that the minds of all, however humble their sphere of service, can be invigorated and cheered. The woman, [Pg 38]who is rearing a family of children; the woman, who labors in the schoolroom; the woman, who, in her retired chamber, earns, with her needle, the mite, which contributes to the intellectual and moral elevation of her Country; even the humble domestic, whose example and influence may be moulding and forming young minds, while her faithful services sustain a prosperous domestic state;—each and all may be animated by the consciousness, that they are agents in accomplishing the greatest work that ever was committed to human responsibility. It is the building of a glorious temple, whose base shall be coextensive with the bounds of the earth, whose summit shall pierce the skies, whose splendor shall beam on all lands; and those who hew the lowliest stone, as much as those who carve the highest capital, will be equally honored, when its top-stone shall be laid, with new rejoicings of the morning stars, and shoutings of the sons of God.

FOOTNOTE:

[A] Miss Martineau is a singular exception to this remark. After receiving unexampled hospitalities and kindnesses, she gives the following picture of her entertainers. Having in other places spoken of the American woman as having "her intellect confined," and "her morals crushed," and as deficient in education, because she has "none of the objects in life for which an enlarged education is considered requisite," she says,—"It is assumed, in America, particularly in New England, that the morals of society there are peculiarly pure. I am grieved to doubt the fact; but I do doubt it." "The Auld-Robin-Gray story is a frequently-enacted tragedy here; and one of the worst symptoms that struck me, was, that there was usually a demand upon my sympathy in such cases." — "The unavoidable consequence of such a mode of marrying, is, that the sanctity of marriage is impaired, and that vice succeeds. There are sad tales in country villages, here and there, that attest this; and yet more in towns, in a rank of society where such things are seldom or never heard of in England." — "I unavoidably knew of more cases of lapse in highly respectable families in one State, than ever came to my knowledge at home; and they were got over with a disgrace far more temporary and superficial than they could have been visited with in England." — "The vacuity of mind of many women, is, I conclude, the cause of a vice, which it is painful to allude to, but which cannot honestly be passed over. — It is no secret on the spot, that the habit of intemperance is not infrequent among women of station and education in the most enlightened parts of the Country. I witnessed some instances, and heard of more. It does not seem to me to be regarded with all the dismay which such a symptom ought to excite. To the stranger, a novelty so horrible, a spectacle so fearful, suggests wide and deep subjects of investigation."

It is not possible for language to give representations more false in every item. In evidence of this, the writer would mention, that, within the last few years, she has travelled almost the entire route taken by Miss Martineau, except the lower tier of the Southern States; and, though not meeting the same individuals, has mingled in the very same circles. Moreover, she has *resided* from several months to several years in *eight* of the different Northern and Western States, and spent several weeks at a time in five other States. She has also had pupils from every State in the Union, but two, and has visited extensively at their houses. But in her whole life, and in all these different positions, the writer has never, to her knowledge, seen even *one* woman, of the classes with which she has associated, who had lapsed in the manner indicated by Miss Martineau; nor does she believe that such a woman could find admission in such circles any where in the Country. As to intemperate women, *five* cases are all of whom the writer has ever heard, in such circles, and two of these many believed to be unwarrantably suspected. After following in Miss Martineau's track, and discovering all the falsehood, twaddle, gossip, old saws, and almanac stories, which have been strung together in her books, no charitable mode of accounting for the medley remains, but to suppose her the pitiable dupe of that love of hoaxing so often found in our Country.

Again, Miss Martineau says, "We passed an unshaded meadow, where the grass had caught fire, *every day*, at *eleven o'clock*, the preceding Summer. This demonstrates the necessity of shade"! A woman, with so little common sense, as to swallow such an absurdity for truth, and then tack to it such an astute deduction, must be a tempting subject for the abovementioned mischievous propensity.

CHAPTER II.
DIFFICULTIES PECULIAR TO AMERICAN WOMEN.

In the preceding chapter, were presented those views, which are calculated to inspire American women with a sense of their high responsibilities to their Country, and to the world; and of the excellence and grandeur of the object to which their energies may be consecrated.

But it will be found to be the law of moral action, that whatever involves great results and great benefits, is always attended with great hazards and difficulties. And as it has been shown, that American women have a loftier position, and a more elevated object of enterprise, than the females of any other nation, so it will appear, that they have greater trials and difficulties to [Pg 39]overcome, than any other women are called to encounter.

Properly to appreciate the nature of these trials, it must be borne in mind, that the estimate of evils and privations depends, not so much on their positive nature, as on the character and habits of the person who meets them. A woman, educated in the savage state, finds it no trial to be destitute of many conveniences, which a woman, even of the lowest condition, in this Country, would deem indispensable to existence. So a woman, educated with the tastes and habits of the best New England or Virginia housekeepers, would encounter many deprivations and trials, which would never occur to one reared in the log cabin of a new settlement. So, also, a woman, who has been accustomed to carry forward her arrangements with well-trained domestics, would meet a thousand trials to her feelings and temper, by the substitution of ignorant foreigners, or shiftless slaves, which would be of little account to one who had never enjoyed any better service.

Now, the larger portion of American women are the descendants of English progenitors, who, as a nation, are distinguished for systematic housekeeping, and for a great love of order, cleanliness, and comfort. And American women, to a greater or less extent, have inherited similar tastes and habits. But the prosperity and democratic tendencies of this Country produce results, materially affecting the comfort of housekeepers, which the females of monarchical and aristocratic lands are not called to meet. In such countries, all ranks and classes are fixed in a given position, and each person is educated for a particular sphere and style of living. And the dwellings, conveniences, and customs of life, remain very nearly the same, from generation to generation. This secures the preparation of all classes for their particular station, and makes the lower orders more dependent, and more subservient to employers.

[Pg 40]

But how different is the state of things in this Country. Every thing is moving and changing. Persons in poverty, are rising to opulence, and persons of wealth, are sinking to poverty. The children of common laborers, by their talents and enterprise, are becoming nobles in intellect, or wealth, or office; while the children of the wealthy, enervated by indulgence, are sinking to humbler stations. The sons of the wealthy are leaving the rich mansions of their fathers, to dwell in the log cabins of the forest, where very soon they bear away the daughters of ease and refinement, to share the privations of a new settlement. Meantime, even in the more stationary portions of the community, there is a mingling of all grades of wealth, intellect, and education. There are no distinct classes, as in aristocratic lands, whose bounds are protected by distinct and impassable lines, but all are thrown into promiscuous masses. Thus, persons of humble means are brought into contact with those of vast wealth, while all intervening grades are placed side by side. Thus, too, there is a constant comparison of conditions, among equals, and a constant temptation presented to imitate the customs, and to strive for the enjoyments, of those who possess larger means.

In addition to this, the flow of wealth, among all classes, is constantly increasing the number of those who live in a style demanding much hired service, while the number of those, who are compelled to go to service, is constantly diminishing. Our manufactories, also, are making increased demands for female labor, and offering larger compensation. In consequence of these things, there is such a disproportion between those who wish to hire, and those who are willing to go to domestic service, that, in the non-slaveholding States, were it not for the supply of poverty-stricken foreigners, there would not be a domestic for each family who demands one. And this resort to foreigners, poor as it is, scarcely meets the demand; while the disproportion must every year increase, especially if our prosperity [Pg 41]increases. For, just in proportion as wealth rolls in upon us, the number of those, who will give up their own independent homes to serve strangers, will be diminished.

The difficulties and sufferings, which have accrued to American women, from this cause, are almost incalculable. There is nothing, which so much demands system and regularity, as the affairs of a housekeeper, made up, as they are, of ten thousand desultory and minute items; and yet, this perpetually fluctuating state of society seems forever to bar any such system and regularity. The anxieties, vexations, perplexities, and even hard labor, which come upon American women, from this state of domestic service, are endless; and many a woman has, in consequence, been disheartened, discouraged, and ruined in health. The only wonder is, that, amid so many real difficulties, American women are still able to maintain such a character for energy, fortitude, and amiableness, as is universally allowed to be their due.

But the second, and still greater difficulty, peculiar to American women, is, a delicacy of constitution, which renders them early victims to disease and decay.

The fact that the women of this Country are unusually subject to disease, and that their beauty and youthfulness are of shorter continuance than those of the women of other nations, is one which always attracts the attention of foreigners; while medical men and philanthropists are constantly giving fearful monitions as to the extent and alarming increase of this evil. Investigations make it evident, that a large proportion of young ladies, from the wealthier classes, have the incipient stages of curvature of the spine, one of the most sure and fruitful causes of future disease and decay. The writer has heard medical men, who have made extensive inquiries, say, that a very large proportion of the young women at boarding schools, are affected in this way, while many other indications of disease and [Pg 42]debility exist, in cases where this particular evil cannot be detected.

In consequence of this enfeebled state of their constitutions, induced by a neglect of their physical education, as soon as they are called to the responsibilities and trials of domestic life, their constitution fails, and their whole existence is rendered a burden. For no woman can enjoy existence, when disease throws a dark cloud over the mind, and incapacitates her for the proper discharge of every duty.

The writer, who for some ten years has had the charge of an institution, consisting of young ladies from almost every State in the Union, since relinquishing that charge, has travelled and visited extensively in most of the non-slaveholding States. In these circuits, she has learned the domestic history, not merely of her pupils, but of many other young wives and mothers, whose sorrowful experience has come to her knowledge. And the impression, produced by the dreadful extent of this evil, has at times been almost overwhelming.

It would seem as if the primeval curse, which has written the doom of pain and sorrow on one period of a young mother's life, in this Country had been extended over all; so that the hour seldom arrives, when "she forgetteth her sorrow for joy that a man is born into the world." Many a mother will testify, with shuddering, that the most exquisite sufferings she ever endured, were not those appointed by Nature, but those, which, for week after week, have worn down health and spirits, when nourishing her child. And medical men teach us, that this, in most cases, results from a debility of constitution, consequent on the mismanagement of early life. And so frequent and so mournful are these, and the other distresses that result from the delicacy of the female constitution, that the writer has repeatedly heard mothers say, that they had wept tears of bitterness over their infant daughters, at the thought of the sufferings which they were destined to undergo; while they cherished [Pg 43]the decided wish, that these daughters should never marry. At the same time, many a reflecting young woman is looking to her future prospects, with very different feelings and hopes from those which Providence designed.

A perfectly healthy woman, especially a perfectly healthy mother, is so unfrequent, in some of the wealthier classes, that those, who are so, may be regarded as the exceptions, and not as the general rule. The writer has heard some of her friends declare, that they would ride fifty miles, to see a perfectly healthy and vigorous woman, out of the laboring classes. This, although somewhat jocose, was not an entirely unfair picture of the true state of female health in the wealthier classes.

There are many causes operating, which serve to perpetuate and increase this evil. It is a well-known fact, that mental excitement tends to weaken the physical system, unless it is counterbalanced by a corresponding increase of exercise and fresh air. Now, the people of this Country are under the influence of high commercial, political, and religious stimulus, altogether greater than was ever known by any other nation; and in all this, women are made the sympathizing companions of the other sex. At the same time, young girls, in pursuing an education, have ten times greater an amount of intellectual taxation demanded, than was ever before exacted. Let any daughter, educated in our best schools at this day, compare the course of her study with that pursued in her mother's early life, and it will be seen that this estimate of the increase of mental taxation probably falls below the truth. Though, in some countries, there are small classes of females, in the higher circles, who pursue literature and science to a far greater extent than in any corresponding circles in this Country, yet, in no nation in the world are the advantages of a good intellectual education enjoyed, by so large a proportion of the females. And this education has consisted far less of accomplishments, and far more of those solid studies which demand the exercise of the [Pg 44]various powers of mind, than the education of the women of other lands.

And when American women are called to the responsibilities of domestic life, the degree in which their minds and feelings are taxed, is altogether greater than it is in any other nation.

No women on earth have a higher sense of their moral and religious responsibilities, or better understand, not only what is demanded of them, as housekeepers, but all the claims that rest upon them as wives, mothers, and members of a social community. An American woman, who is the mistress of a family, feels her obligations, in reference to her influence over her husband, and a still greater responsibility in rearing and educating her children. She feels, too, the claims which the moral interests of her domestics have on her watchful care. In social life, she recognises the claims of hospitality, and the demands of friendly visiting. Her responsibility, in reference to the institutions of benevolence and religion, is deeply realized. The regular worship of the Lord's day, and all the various religious meetings and benevolent societies which place so much dependence on female influence and example, she feels obligated to sustain. Add to these multiplied responsibilities, the perplexities and evils which have been pointed out, resulting from the fluctuating state of society, and the deficiency of domestic service, and no one can deny that American women are exposed to a far greater amount of intellectual and moral excitement, than those of any other land. Of course, in order to escape the danger resulting from this, a greater amount of exercise in the fresh air, and all those methods which strengthen the constitution, are imperiously required.

But, instead of this, it will be found, that, owing to the climate and customs of this Nation, there are no women who secure so little of this healthful and protecting regimen, as ours. Walking and riding and gardening, in the open air, are practised by the women of other lands, to a far greater extent, than by American females. [Pg 45]Most English women, in the wealthier classes, are able to walk six and eight miles, without oppressive fatigue; and when they visit this Country, always express their surprise at the inactive habits of American ladies. In England, regular exercise, in the open air, is very commonly required by the mother, as a part of daily duty, and is sought by young women, as an enjoyment. In consequence of a different physical training, English women, in those circles which enjoy competency, present an appearance which always strikes American gentlemen as a contrast to what they see at home. An English mother, at thirty, or thirty-five, is in the full bloom of perfected womanhood; as fresh and healthful as her daughters. But where are the American mothers, who can reach this period unfaded and unworn? In America, young ladies of the wealthier classes are sent to school from early childhood; and neither parents nor teachers make it a definite object to secure a proper amount of fresh air and exercise, to counterbalance this intellectual taxation. As soon as their school days are over, dressing, visiting, evening parties, and stimulating amusements, take the place of study, while the most unhealthful modes of dress add to the physical exposures. To make morning calls, or do a little shopping, is all that can be termed their exercise in the fresh air; and this, compared to what is needed, is absolutely nothing, and on some accounts is worse than nothing.[B] In consequence of these, and other evils, which will be pointed out more at large in the following pages, the young women of America grow up with such a delicacy of constitution, that probably eight out of ten become subjects of disease, either before or as soon as they are called to the responsibilities of domestic life.

But there is one peculiarity of situation, in regard to American women, which makes this delicacy of constitution [Pg 46]still more disastrous. It is the liability to the exposures and hardships of a newly-settled country.

One more extract from De Tocqueville will give a view of this part of the subject, which any one, familiar with Western life, will admire for its verisimilitude.

"The same strength of purpose which the young wives of America display in bending themselves, at once, and without repining, to the austere duties of their new condition, is no less manifest in all the great trials of their lives. In no country in the world, are private fortunes more precarious, than in the United States. It is not uncommon for the same man, in the course of his life, to rise and sink again through all the grades which lead from opulence to poverty. American women support these vicissitudes with a calm and unquenchable energy. It would seem that their desires contract, as easily as they expand, with their fortunes. The greater part of the adventurers, who migrate, every year, to people the Western wilds, belong" "to the old Anglo-American race of the Northern States. Many of these men, who rush so boldly onward in pursuit of wealth, were already in the enjoyment of a competency in their own part of the Country. They take their wives along with them, and make them share the countless perils and privations, which always attend the commencement of these expeditions. I have often met, even on the verge of the wilderness, with young women, who, after having been brought up amid all the comforts of the large towns of New England, had passed, almost without any intermediate stage, from the wealthy abode of their parents, to a comfortless hovel in a forest. Fever, solitude, and a tedious life, had not broken the springs of their courage. Their features were impaired and faded, but their looks were firm: they appeared to be, at once, sad and resolute."

In another passage, he gives this picturesque sketch: "By the side of the hearth, sits a woman, with a baby on her lap. She nods to us, without disturbing herself. Like the pioneer, this woman is in the prime of life; [Pg 47]her appearance would seem superior to her condition: and her apparel even betrays a lingering taste for dress. But her delicate limbs appear shrunken; her features are drawn in; her eye is mild and melancholy; her whole physiognomy bears marks of a degree of religious resignation, a deep quiet of all passion, and some sort of natural and tranquil firmness, ready to meet all the ills of life, without fearing and without braving them. Her children cluster about her, full of health, turbulence, and energy; they are true children of the wilderness: their mother watches them, from time to time, with mingled melancholy and joy. To look at their strength, and her languor, one might imagine that the life she had given them had exhausted her own; and still she regrets not what they have cost her. The house, inhabited by these emigrants, has no internal partition or loft. In the one chamber of which it consists, the whole family is gathered for the night. The dwelling is itself a little world; an ark of civilization amid an ocean of foliage. A hundred steps beyond it, the primeval forest spreads its shades, and solitude resumes its sway."

Such scenes, and such women, the writer has met, and few persons realize how many refined and lovely women are scattered over the broad prairies and deep forests of the West; and none, but the Father above, appreciates the extent of those sacrifices and sufferings, and the value of that firm faith and religious hope, which live, in perennial bloom, amid those vast solitudes. If the American women of the East merit the palm, for their skill and success as accomplished housekeepers, still more is due to the heroines of the West, who, with such unyielding fortitude and cheerful endurance, attempt similar duties, amid so many disadvantages and deprivations.

But, though American women have those elevated principles and feelings, which enable them to meet such trials in so exemplary a manner, their physical energies are not equal to the exertions demanded. Though the [Pg 48]mind may be bright and firm, the casket is shivered; though the spirit may be willing, the flesh is weak. A woman of firm health, with the hope and elasticity of youth, may be envied rather than pitied, as she shares with her young husband the hopes and enterprises of pioneer life. But, when the body fails, then the eye of hope grows dim, the heart sickens, the courage dies; and, in solitude, weariness, and suffering, the wanderer pines for the dear voices and the tender sympathies of a far distant home. Then it is, that the darkest shade is presented, which marks the peculiar trials and liabilities of American women, and which exhibits still more forcibly the disastrous results of that delicacy of constitution which has been pointed out. For, though all American women, or even the greater part of them, are not called to encounter such trials, yet no mother, who rears a family of daughters, can say, that such a lot will not fall to one of her flock; nor can she know which will escape. The reverses of fortune, and the chances of matrimony, expose every woman in the Nation to such liabilities, for which she needs to be prepared.

FOOTNOTE:

[B] So little idea have most ladies, in the wealthier classes, of what is a proper amount of exercise, that, if they should succeed in walking a mile or so, at a moderate pace, three or four times a week, they would call it taking a great deal of exercise.

CHAPTER III.
REMEDIES FOR THE PRECEDING DIFFICULTIES.

Having pointed out the peculiar responsibilities of American women, and the peculiar embarrassments which they are called to encounter, the following suggestions are offered, as remedies for such difficulties.

In the first place, the physical and domestic education of daughters should occupy the principal attention of mothers, in childhood; and the stimulation of the intellect should be very much reduced. As a general rule, daughters should not be sent to school before they are six years old; and, when they are sent, far more attention should be paid to their physical developement, [Pg 49]than is usually done. They should never be confined, at any employment, more than an hour at a time; and this confinement should be followed by sports in the open air. Such accommodations should be secured, that, at all seasons, and in all weathers, the teacher can every half hour send out a portion of her school, for sports. And still more care should be given to preserve pure air in the schoolroom. The close stoves, crowded condition, and poisonous air, of most schoolrooms, act as constant drains on the health and strength of young children.

In addition to this, much less time should be given to school, and much more to domestic employments, especially in the wealthier classes. A little girl may begin, at five or six years of age, to assist her mother; and, if properly trained, by the time she is ten, she can render essential aid. From this time, until she is fourteen or fifteen, it should be the principal object of her education to secure a strong and healthy constitution, and a thorough practical knowledge of all kinds of domestic employments. During this period, though some attention ought to be paid to intellectual culture, it ought to be made altogether secondary in importance; and such a measure of study and intellectual excitement, as is now demanded in our best female seminaries, ought never to be allowed, until a young lady has passed the most critical period of her youth, and has a vigorous and healthful constitution fully established. The plan might be adopted, of having schools for young girls kept only in the afternoon; that their mornings might be occupied in domestic exercise, without interfering with school employments. Where a proper supply of domestic exercise cannot be afforded, the cultivation of flowers and fruits might be resorted to, as a delightful and unfailing promotive of pleasure and health.

And it is to that class of mothers, who have the best means of securing hired service, and who are the most tempted to allow their daughters to grow up with inactive habits, that their Country and the world must [Pg 50]look for a reformation, in this respect. Whatever ladies in the wealthier classes decide shall be fashionable, will be followed by all the rest; but, while they persist in the aristocratic habits, now so common, and bring up their daughters to feel as if labor was degrading and unbecoming, the evils pointed out will never find a remedy. It is, therefore, the peculiar duty of ladies, who have wealth, to set a proper example, in this particular, and make it their first aim to secure a strong and healthful constitution for their daughters, by active domestic employments. All the sweeping, dusting, care of furniture and beds, the clear starching, and the nice cooking, should be done by the daughters of a family, and not by hired servants. It may cost the mother more care, and she may find it needful to hire a person for the express purpose of instructing and superintending her daughters, in these employments; but it should be regarded as indispensable to be secured, either by the mother's agency, or by a substitute.

It is in this point of view, that the dearth of good domestics in this Country may, in its results, prove a substantial blessing. If all housekeepers, who have the means, could secure good servants, there would be little hope that so important a revolution, in the domestic customs of the wealthy classes, could be effected. And so great is the natural indolence of mankind, that the amount of exercise, needful for health, will never be secured by those who are led to it through no necessity, but merely from rational considerations. Yet the pressure of domestic troubles, from the want of good domestics, has already determined many a mother, in the wealthy classes, to train her daughters to aid her in domestic service; and thus necessity is compelling mothers to do what abstract principles of expediency could never secure.

A second method of promoting the same object, is, to raise the science and practice of Domestic Economy to its appropriate place, as a regular study in female seminaries. The succeeding chapter will present the [Pg 51]reasons for this, more at large. But it is to the mothers of our Country, that the community must look for this change. It cannot be expected, that teachers, who have their attention chiefly absorbed by the intellectual and moral interests of their pupils, should properly realize the importance of this department of education. But if mothers generally become convinced of this, their judgement and wishes will meet the respectful consideration they deserve, and the object will be accomplished.

The third method of securing a remedy for the evils pointed out, is, the endowment of female institutions, under the care of suitable trustees, who shall secure a proper course of education. The importance of this measure cannot be realized by those, who have not turned their attention to this subject; and for such, the following considerations are presented.

The endowment of colleges, and of law, medical, and divinity, schools, for the other sex, is designed to secure a thorough and proper education, for those who have the most important duties of society to perform. The men who are to expound the laws, the men who have the care of the public health, and the men who are to communicate religious instruction, should have well-disciplined and well-informed minds; and it is mainly for this object that collegiate and professional institutions are established. Liberal and wealthy individuals contribute funds, and the legislatures of the States also lend assistance, so that every State in this Nation has from one to twenty such endowed institutions, supplied with buildings, apparatus, a library, and a faculty of learned men to carry forward a superior course of instruction. And the use of all these advantages is secured, in many cases, at an expense, no greater than is required to send a boy to a common school and pay his board there. No private school could offer these advantages, without charging such a sum, as would forbid all but the rich from securing its benefits. By furnishing such superior advantages, on low terms, multitudes are properly educated, who would otherwise [Pg 52]remain in ignorance; and thus the professions are supplied, by men properly qualified for them.

Were there no such institutions, and no regular and appropriate course of study demanded for admission to the bar, the pulpit, and to medical practice, the education of most professional men would be desultory, imperfect, and deficient. Parents and children would regulate the course of study according to their own crude notions; and, instead of having institutions which agree in carrying on a similar course of study, each school would have its own peculiar system, and compete and conflict with every other. Meantime, the public would have no means of deciding which was best, nor any opportunity for learning when a professional man was properly qualified for his duties. But as it is, the diploma of a college, and the license of an appointed body of judges, must both be secured, before a young man feels that he has entered the most promising path to success in his profession.

Our Country, then, is most abundantly supplied with endowed institutions, which secure a liberal education, on such low terms as make them accessible to all classes, and in which the interests of education are watched over, sustained, and made permanent, by an appropriate board of trustees.

But are not the most responsible of all duties committed to the charge of woman? Is it not her profession to take care of mind, body, and soul? and that, too, at the most critical of all periods of existence? And is it not as much a matter of public concern, that she should be properly qualified for her duties, as that ministers, lawyers, and physicians, should be prepared for theirs? And is it not as important, to endow institutions which shall make a superior education accessible to all classes,—for females, as for the other sex? And is it not equally important, that institutions for females be under the supervision of intelligent and responsible trustees, whose duty it shall be to secure a uniform and appropriate education for one sex [Pg 53]as much as for the other? It would seem as if every mind must accord an affirmative reply, as soon as the matter is fairly considered.

As the education of females is now conducted, any man or woman who pleases, can establish a female seminary, and secure recommendations which will attract pupils. But whose business is it to see that these young females are not huddled into crowded rooms?

or that they do not sleep in ill-ventilated chambers? or that they have healthful food? or that they have the requisite amount of fresh air and exercise? or that they pursue an appropriate and systematic course of study? or that their manners, principles, and morals, are properly regulated? Parents either have not the means, or else are not qualified to judge; or, if they are furnished with means and capacity, they are often restricted to a choice of the best school within reach, even when it is known to be exceedingly objectionable.

If the writer were to disclose all that can truly be told of boarding-school life, and its influence on health, manners, disposition, intellect, and morals, the disclosure would both astonish and shock every rational mind. And yet she believes that such institutions are far better managed in this Country, than in any other; and that the number of those, which are subject to imputations in these respects, is much less than could reasonably be expected. But it is most surely the case, that much remains to be done, in order to supply such institutions as are needed for the proper education of American women.

In attempting a sketch of the kind of institutions which are demanded, it is very fortunate that there is no necessity for presenting a theory, which may, or may not, be approved by experience. It is the greatest honor of one of our newest Western States, that it can boast of such an Institution, endowed, too, wholly by the munificence of a single individual. A slight sketch of this Institution, which the writer has examined in all [Pg 54]its details, will give an idea of what can be done, by showing what has actually been accomplished.

This Institution[C] is under the supervision of a Board of Trustees, who hold the property in trust for the object to which it is devoted, and who have the power to fill their own vacancies. It is furnished with a noble and tasteful building, of stone, so liberal in dimensions and arrangement, that it can accommodate ninety pupils and teachers, giving one room to every two pupils, and all being so arranged, as to admit of thorough ventilation. This building is surrounded by extensive grounds, enclosed with handsome fences, where remains of the primeval forest still offer refreshing shade for juvenile sports.

To secure adequate exercise for the pupils, two methods are adopted. By the first, each young lady is required to spend a certain portion of time in domestic employments, either in sweeping, dusting, setting and clearing tables, washing and ironing, or other household concerns.

Let not the aristocratic mother and daughter express their dislike of such an arrangement, till they can learn how well it succeeds. Let them walk, as the writer has done, through the large airy halls, kept clean and in order by their fair occupants, to the washing and ironing-rooms. There they will see a long hall, conveniently fitted up with some thirty neatly-painted tubs, with a clean floor, and water conducted so as to save both labor and slopping. Let them see some thirty or forty merry girls, superintended by a motherly lady, chatting and singing, washing and starching, while every convenience is at hand, and every thing around is clean and comfortable. Two hours, thus employed, enable[Pg 55] each young lady to wash the articles she used during the previous week, which is all that is demanded, while thus they are all practically initiated into the arts and mysteries of the wash-tub. The Superintendent remarked to the writer, that, after a few weeks of probation, most of her young washers succeeded quite as well as those whom she could hire, and who made it their business. Adjacent to the washing-room, is the ironing establishment; where another class are arranged, on the ironing-day, around long, extended tables, with heating-furnaces, clothes-frames, and all needful appliances.

By a systematic arrangement of school and domestic duties, a moderate portion of time, usually not exceeding two hours a day, from each of the pupils, accomplished all the domestic labor of a family of ninety, except the cooking, which was done by two hired domestics. This part of domestic labor it was deemed inexpedient to incorporate as a portion of the business of the pupils, inasmuch as it could not be accommodated to the arrangements of the school, and was in other respects objectionable.

Is it asked, how can young ladies paint, play the piano, and study, when their hands and dresses must be unfitted by such drudgery? The woman who asks this question, has yet to learn that a pure and delicate skin is better secured by healthful exercise, than by any other method; and that a young lady, who will spend two hours a day at the wash-tub, or with a broom, is far more likely to have rosy cheeks, a finely-moulded form, and a delicate skin, than one who lolls all day in her parlor or chamber, or only leaves it, girt in tight dresses, to make fashionable calls. It is true, that long-protracted daily labor hardens the hand, and unfits it for delicate employments; but the amount of labor needful for health produces no such effect. As to dress, and appearance, if neat and convenient accommodations are furnished, there is no occasion for the exposures which demand shabby dresses. A dark calico, [Pg 56]genteelly made, with an oiled-silk apron, and wide cuffs of the same material, secures both good looks and good service. This plan of domestic employments for the pupils in this Institution, not only secures regular healthful exercise, but also aids to reduce the expenses of education, so that, with the help of the endowments, it is brought within the reach of many, who otherwise could never gain such advantages.

In addition to this, a system of Calisthenic[D] exercises is introduced, which secures all the advantages which dancing is supposed to effect, and which is free from the dangerous tendencies of that fascinating and fashionable amusement. This system is so combined with music, and constantly varying evolutions, as to serve as an amusement, and also as a mode of curing distortions, particularly all tendencies to curvature of the spine; while, at the same time, it tends to promote grace of movement, and easy manners.

Another advantage of this Institution, is, an elevated and invigorating course of mental discipline. Many persons seem to suppose, that the chief object of an intellectual education is the acquisition of knowledge. But it will be found, that this is only a secondary object. The formation of habits of investigation, of correct reasoning, of persevering attention, of regular system, of accurate analysis, and of vigorous mental action, is the primary object to be sought in preparing American women for their arduous duties; duties which will demand not only quickness of perception, but steadiness of purpose, regularity of system, and perseverance in action.

It is for such purposes, that the discipline of the Mathematics is so important an element in female [Pg 57]education; and it is in this aspect, that the mere acquisition of facts, and the attainment of ac-complishments, should be made of altogether secondary account.

In the Institution here described, a systematic course of study is adopted, as in our colleges; designed to occupy three years. The following slight outline of the course, will exhibit the liberal plan adopted in this respect.

In Mathematics, the whole of Arithmetic contained in the larger works used in schools, the whole of Euclid, and such portions from Day's Mathematics as are requisite to enable the pupils to demonstrate the various problems in Olmsted's larger work on Natural Philosophy. In Language, besides English Grammar, a short course in Latin is required, sufficient to secure an understanding of the philosophy of the language, and that kind of mental discipline which the exercise of translating affords. In Philosophy, Chemistry, Astronomy, Botany, Geology and Mineralogy, Intellectual and Moral Philosophy, Political Economy, and the Evidences of Christianity, the same textbooks are used as are required at our best colleges. In Geography, the most thorough course is adopted; and in History, a more complete knowledge is secured, by means of charts and textbooks, than most of our colleges offer. To these branches, are added Griscom's Physiology,[E] Bigelow's Technology, and Jahn's Archæology, together with a course of instruction in polite literature, for which Chambers's English Literature is employed as the text-book, each recitation being attended with selections and criticisms, from teacher or pupils, on the various authors brought into notice. Vocal Music, on the plan of the Boston Academy, is a part of the daily instructions. [Pg 58]Linear drawing, and pencilling, are designed also to be a part of the course. Instrumental Music is taught, but not as a part of the regular course of study.

To secure the proper instruction in all these branches, the division of labor, adopted in colleges, is pursued. Each teacher has distinct branches as her department, for which she is responsible, and in which she is independent. One teacher performs the duties of a *governess*, in maintaining rules, and attending to the habits and manners of the pupils. By this method, the teachers have sufficient time, both to prepare themselves, and to impart instruction and illustration in the class-room. In this Institution it is made a direct object of effort *to cure defects* of *character and habits*. At the frequent meetings of the Principal and teachers, the peculiarities of each pupil are made the subjects of inquiry; and methods are devised for remedying defects through the personal influence of the several teachers. This, when thus made a direct object of combined effort, often secures results most gratifying and encouraging.

One peculiarity of this Institution demands consideration. By the method adopted here, the exclusive business of educating their own sex is, as it ever ought to be, confined to females. The Principal of the Institution, indeed, is a gentleman; but, while he takes the position of a father of the family, and responsible head of the whole concern, the entire charge of instruction, and most of the responsibilities in regard to health, morals, and manners, rest upon the female teachers, in their several departments. The Principal is the chaplain and religious teacher; and is a member of the board of instructors, so far as to have a right to advise, and an equal vote, in every question pertaining to the concerns of the School; and thus he acts as a sort of regulator and mainspring in all the various departments. But no one person in the Institution is loaded with the excessive responsibilities, which rest upon one, where a large institution of this kind has a Principal, who employs and directs all the subordinate [Pg 59]assistants. The writer has never before seen the principle of the division of labor and responsibility so perfectly carried out in any female institution; and she believes that experience will prove that this is the true model for combining, in appropriate proportions, the agency of both sexes in carrying forward such an institution. There are cases where females are well qualified, and feel willing to take the place occupied by the Principal; but such cases are rare.

One thing more should be noticed, to the credit of the rising State where this Institution is located. A female association has been formed, embracing a large portion of the ladies of standing and wealth, the design of which, is, to educate, gratuitously, at this, and other similar, institutions, such females as are anxious to obtain a good education, and are destitute of the means. If this enterprise is continued, with the same energy and perseverance as has been manifested during the last few years, that State will take the lead of her sister States in well-educated women; and if the views in the preceding pages are correct, this will give her precedence in every intellectual and moral advantage.

Many, who are not aware of the great economy secured by a proper division of labor, will not understand how so extensive a course can be properly completed in three years. But in this Institution, none are received under fourteen; and a certain amount of previous acquisition is required, in order to admission, as is done in our colleges. This secures a diminution of classes, so that but few studies are pursued at one time; while the number of well-qualified teachers is so adequate, that full time is afforded for all needful instruction and illustration. Where teachers have so many classes, that they merely have time to find out what the pupils learn from books, without any aid from their teachers, the acquisitions of the pupils are vague and imperfect, and soon pass away; so that an immense amount of expense, time, and labor, is spent in acquiring or recalling what is lost about as fast as it is gained.

[Pg 60]

Parents are little aware of the immense waste incurred by the present mode of conducting female education. In the wealthy classes, young girls are sent to school, as a matter of course, year after year, confined, for six hours a day, to the schoolhouse, and required to add some time out of school to learning their lessons. Thus, during the most critical period of life, they are for a long time immured in a room, filled with an atmosphere vitiated by many breaths, and are constantly kept under some sort of responsibility in regard to mental effort. Their studies are pursued at random, often changed with changing schools, while book after book (heavily taxing the parent's purse) is conned awhile, and then supplanted by others. Teachers have usually so many pupils, and such a variety of branches to teach, that little time can be afforded to each pupil; while scholars, at this thoughtless period of life, feeling sure of going to school as long as they please, manifest little interest in their pursuits.

The writer believes that the actual amount of education, permanently secured by most young ladies from the age of ten to fourteen, could all be acquired in one year, at the Institution described, by a young lady at the age of fifteen or sixteen.

Instead of such a course as the common one, if mothers would keep their daughters as their domestic assistants, until they are fourteen, requiring them to study one lesson, and go out, once a day, to recite it to a teacher, it would abundantly prepare them, after their constitutions are firmly established, to enter such an institution, where, in three years, they could secure more, than almost any young lady in the Country now gains by giving the whole of her youth to school pursuits.

In the early years of female life, reading, writing, needlework, drawing, and music, should alternate with domestic duties; and one hour a day, devoted to some study, in addition to the above pursuits, would be all that is needful to prepare them for a thorough education [Pg 61]after growth is attained, and the constitution established. This is the time when young women would feel the value of an education, and pursue their studies with that maturity of mind, and vividness of interest, which would double the perpetuity and value of all their acquisitions.

The great difficulty, which opposes such a plan, is, the want of institutions that would enable a young lady to complete, in three years, the liberal course of study, here described. But if American mothers become convinced of the importance of such advantages for their daughters, and will use their influence appropriately and efficiently, they will certainly be furnished. There are other men of liberality and wealth, besides the individual referred to, who can be made to feel that a fortune, expended in securing an appropriate education to American women, is as wisely bestowed, as in founding colleges for the other sex, who are already so abundantly supplied. We ought to have institutions, similar to the one described, in every part of this Nation; and funds should be provided, for educating young women destitute of means: and if American women think and feel, that, by such a method, their own trials will be lightened, and their daughters will secure a healthful constitution and a thorough domestic and intellectual education, the appropriate expression of their wishes will secure the necessary funds. The tide of charity, which has been so long flowing from the female hand to provide a liberal education for young men, will flow back with abundant remuneration.

The last method suggested for lessening the evils peculiar to American women, is, a decided effort to oppose the aristocratic feeling, that labor is degrading; and to bring about the impression, that it is refined and lady-like to engage in domestic pursuits. In past ages, and in aristocratic countries, leisure and indolence and frivolous pursuits have been deemed lady-like and refined, because those classes, which were most refined, countenanced such an opinion. But whenever ladies of refinement, [Pg 62]as a general custom, patronise domestic pursuits, then these employments will be deemed lady-like. It may be urged, however, that it is impossible for a woman who cooks, washes, and sweeps, to appear in the dress, or acquire the habits and manners, of a lady; that the drudgery of the kitchen is dirty work, and that no one can appear delicate and refined, while engaged in it. Now all this depends on circumstances. If a woman has a house, destitute of neat and convenient facilities; if she has no habits of order and system; if she is remiss and careless in person and dress;—then all this may be true. But, if a woman will make some sacrifices of costly ornaments in her parlor, in order to make her kitchen neat and tasteful; if she will sacrifice expensive dishes, in order to secure such conveniences for labor as protect from exposures; if she will take pains to have the dresses, in which she works, made of suitable materials, and in good taste; if she will rise early, and systematize and oversee the work of her family, so as to have it done thoroughly, neatly, and in the early part of the day; she will find no necessity for any such apprehensions. It is because such work has generally been done by vulgar people, and in a vulgar way, that we have such associations; and when ladies manage such things, as ladies should, then such associations will be removed. There are pursuits, deemed very refined and genteel, which involve quite as much exposure as kitchen employments. For example, to draw a large landscape, in colored crayons, would be deemed very lady-like; but the writer can testify, from sad experience, that no cooking, washing, sweeping, or any other domestic duty, ever left such deplorable traces on hands, face, and dress, as this same lady-like pursuit. Such things depend entirely on custom and associations; and every American woman, who values the institutions of her Country, and wishes to lend her influence in extending and perpetuating such blessings, may feel that she is doing this, whenever, by her example and influence, she destroys the aristocratic association, which would render domestic labor degrading.

FOOTNOTES:

[C] The writer omits the name of this Institution, lest an inference should be drawn which would be unjust to other institutions. There are others equally worthy of notice, and the writer selects this only because her attention was especially directed to it as being in a new State, and endowed wholly by an individual.

[D] From two Greek words,—καλος, *kalos*, beauty, and σθενος, *sthenos*, strength, being the union of both. The writer is now preparing for the press, an improved system, of her own invention, which, in *some* of its parts, has been successfully introduced into several female seminaries, with advantage. This plan combines singing with a great variety of amusing and graceful evolutions, designed to promote both health and easy manners.

[E] This work, which has gone through numerous editions, and been received by the public with great favour, forms No. lxxxv. of the "Family Library," and No. lvii. of the "School District Library," issued by the publishers of this volume. It is abundantly illustrated by engravings, and has been extensively introduced as a school text-book.

[Pg 63]

CHAPTER IV.
ON DOMESTIC ECONOMY AS A BRANCH OF STUDY.

The greatest impediment to making Domestic Economy a branch of study, is, the fact, that neither parents nor teachers realize the importance, or the practicability of constituting it a regular part of school education.

It is with reference to this, that the first aim of the writer will be, to point out some of the reasons for introducing Domestic Economy as a branch of female education, to be studied at school.

The first reason, is, that there is no period, in a young lady's life, when she will not find such knowledge useful to herself and to others. The state of domestic service, in this Country, is so precarious, that there is scarcely a family, in the free States, of whom it can be affirmed, that neither sickness, discontent, nor love of change, will deprive them of all their domestics, so that every female member of the family will be required to lend some aid, in providing food and the conveniences of living; and the better she is qualified to render it, the happier she will be, and the more she will contribute to the enjoyment of others.

A second reason, is, that every young lady, at the close of her schooldays, and even before they are closed, is liable to be placed in a situation, in which she will need to do, herself, or to teach others to do, all the various processes and duties detailed in this work. That this may be more fully realized, the writer will detail some instances, which have come under her own observation.

The eldest daughter of a family returned from school, on a visit, at sixteen years of age. Before her vacation had closed, her mother was laid in the grave; and such were her father's circumstances, that she was obliged to assume the cares and duties of her lost parent. The [Pg 64]care of an infant, the management of young children, the superintendence of domestics, the charge of family expenses, the responsibility of entertaining company, and the many other cares of the family state, all at once came upon this young and inexperienced schoolgirl.

Again; a young lady went to reside with a married sister, in a distant State. While on this visit, the elder sister died, and there was no one but this young lady to fill the vacant place, and assume all the cares of the nursery, parlor, and kitchen.

Again; a pupil of the writer, at the end of her schooldays, married, and removed to the West. She was an entire novice in all domestic matters; an utter stranger in the place to which she removed. In a year, she became a mother, and *her health failed*; while, for most of the time, she had no domestics, at all, or only Irish or Germans, who scarcely knew even the names, or the uses, of many cooking utensils. She was treated with politeness by her neighbors, and wished to return their civilities; but how could this young and delicate creature, who had spent all her life at school, or in visiting and amusement, take care of her infant, attend to her cooking, washing, ironing, and baking, the concerns of her parlor, chambers, kitchen, and cellar, and yet visit and receive company? If there is any thing that would make a kindly heart ache, with sorrow and sympathy, it would be to see so young, so amiable, so helpless a martyr to the mistaken system of female education now prevalent. "I have the kindest of husbands," said the young wife, after her narrative of sufferings, "and I never regretted my marriage; but, since this babe was born, I have never had a single waking hour of freedom from anxiety and care. O! how little young girls know what is before them, when they enter married life!" Let the mother or teacher, whose eye may rest on these lines, ask herself, if there is no cause for fear that the young objects of her care may be thrown into similar emergencies, where they may need a kind of preparation, which as yet has been withheld.

[Pg 65]

Another reason for introducing such a subject, as a distinct branch of school education, is, that, as a general fact, young ladies *will not* be taught these things in any other way. In reply to the thousand-times-repeated remark, that girls must be taught their domestic duties by their mothers, at home, it may be inquired, in the first place, What proportion of mothers are qualified to teach a *proper* and *complete* system of Domestic Economy? When this is answered, it may be asked, What proportion of those who are qualified, have that sense of the importance of such instructions, and that energy and perseverance which would enable them actually to teach their daughters, in all the branches of Domestic Economy presented in this work?

It may then be asked, How many mothers *actually do* give their daughters instruction in the various branches of Domestic Economy? Is it not the case, that, owing to ill health, deficiency of domestics, and multiplied cares and perplexities, a large portion of the most intelligent mothers, and those, too, who most realize the importance of this instruction, actually cannot find the time, and have not the energy, necessary to properly perform the duty? They are taxed to the full amount of both their mental and physical energies, and cannot attempt any thing more. Almost every woman knows, that it is easier to do the work, herself, than it is to teach an awkward and careless novice; and the great majority of women, in this Country, are obliged to do almost every thing in the shortest and easiest way. This is one reason why the daughters of very energetic and accomplished housekeepers are often the most deficient in these respects; while the daughters of ignorant or inefficient mothers, driven to the exercise of their own energies, often become the most systematic and expert.

It may be objected, that such things cannot be taught by books. This position may fairly be questioned. Do not young ladies learn, from books, how to make hydrogen and oxygen? Do they not have pictures of furnaces, [Pg 66]alembics, and the various utensils employed in *cooking* the chemical agents? Do they not study the various processes of mechanics, and learn to understand and to do many as difficult operations, as any that belong to housekeeping? All these things are explained, studied, and recited in classes, when every one knows that little practical use can ever be made of this knowledge. Why, then, should not that science and art, which a woman is to practise during her whole life, be studied and recited?

It may be urged, that, even if it is studied, it will soon be forgotten. And so will much of every thing studied at school. But why should that knowledge, most needful for daily comfort, most liable to be in demand, be the only study omitted, because it may be forgotten?

It may also be objected, that young ladies can get such books, and attend to them out of school. And so they can get books on Chemistry and Philosophy, and study them out of school; but *will* they do it? And why ought we not to make sure of the most necessary knowledge, and let the less needful be omitted? If young ladies study such a work as this, in school, they will remember a great part of it; and, when they forget, in any emergency, they will know where to resort for instruction. But if such books are not put into schools, probably not one in twenty will see or hear of them, especially in those retired places where they are most needed. And is it at all probable, that a branch, which is so lightly esteemed as to be deemed unworthy a place in the list of female studies, will be sought for and learned by young girls, who so seldom look into works of solid instruction after they leave school? So deeply is the writer impressed with the importance of this, as a branch of female education, at school, that she would deem it far safer and wiser to omit any other, rather than this.

Another reason, for introducing such a branch of study into female schools, is, the influence it would exert, [Pg 67]in leading young ladies more correctly to estimate the importance and dignity of domestic knowledge. It is now often the case, that young ladies rather pride themselves on their ignorance of such subjects; and seem to imagine that it is vulgar and ungenteel to know how to work. This is one of the relics of an aristocratic state of society, which is fast passing away. Here, the tendency of every thing is to the equalisation of labor, so that all classes are feeling, more and more, that indolence is disreputable. And there are many mothers, among the best educated and most wealthy classes, who are bringing up their daughters, not only to know how to do, but actually to do, all kinds of domestic work. The writer knows young ladies, who are daughters of men of wealth and standing, and who are among the most accomplished in their sphere, who have for months been sent to work with a mantuamaker, to acquire a practical knowledge of her occupation, and who have at home learned to perform all kinds of domestic labor.

And let the young women of this Nation find, that Domestic Economy is placed, in schools, on equal or superior ground to Chemistry, Philosophy, and Mathematics, and they will blush to be found ignorant of its first principles, as much as they will to hesitate respecting the laws of gravity, or the composition of the atmosphere. But, as matters are now conducted, many young ladies know how to make oxygen and hydrogen, and to discuss questions of Philosophy or Political Economy, far better than they know how to make a bed and sweep a room properly; and they can "construct a diagram" in Geometry, with far more skill than they can make the simplest article of female dress.

It may be urged, that the plan suggested by the writer, in the previous pages, would make such a book as this needless; for young ladies would learn all these things at home, before they go to school. But it must be remembered, that the plan suggested cannot fully be carried into effect, till such endowed institutions, as the [Pg 68]one described, are universally furnished. This probably will not be done, till at least one generation of young women are educated. It is only on the supposition that a young lady can, at fourteen or fifteen years of age, enter such an institution, and continue there three years, that it would be easy to induce her to remain, during all the previous period, at home, in the practice of Domestic Economy, and the limited course of study pointed out. In the present imperfect, desultory, varying, mode of female education, where studies are begun, changed, partially learned, and forgotten, it requires nearly all the years of a woman's youth, to acquire the intellectual education now demanded. While this state of things continues, the only remedy is, to introduce Domestic Economy as a study at school.

It is hoped that these considerations will have weight, not only with parents and teachers, but with young ladies themselves, and that all will unite their influence to introduce this, as a popular and universal branch of education, into every female school.

CHAPTER V.
ON THE CARE OF HEALTH.

There is no point, where a woman is more liable to suffer from a want of knowledge and experience, than in reference to the health of a family committed to her care. Many a young lady, who never had any charge of the sick; who never took any care of an infant; who never obtained information on these subjects from books, or from the experience of others; in short, with little or no preparation; has found herself the principal attendant in dangerous sickness, the chief nurse of a feeble infant, and the responsible guardian of the health of a whole family.

The care, the fear, the perplexity, of a woman, suddenly [Pg 69]called to these unwonted duties, none can realize, till they themselves feel it, or till they see some young and anxious novice first attempting to meet such responsibilities. To a woman of age and experience, these duties often involve a measure of trial and difficulty, at times deemed almost insupportable; how hard, then, must they press on the heart of the young and inexperienced!

There is no really efficacious mode of preparing a woman to take a *rational* care of the health of a family, except by communicating that knowledge, in regard to the construction of the body, and the laws of health, which is the basis of the medical profession. Not that a woman should undertake the minute and extensive investigation requisite for a physician; but she should gain a general knowledge of first principles, as a guide to her judgement in emergencies when she can rely on no other aid. Therefore, before attempting to give any specific directions on the subject of this chapter, a short sketch of the construction of the human frame will be given, with a notice of some of the general principles, on which specific rules in regard to health are based. This description will be arranged under the general heads of Bones, Muscles, Nerves, Blood-Vessels, Organs of Digestion and Respiration, and the Skin.

BONES.

The bones are the most solid parts of the body. They are designed to protect and sustain it, and also to secure voluntary motion. They are about two hundred and fifty in number, (there being sometimes a few more or less,) and are fastened together by cartilage, or gristle, a substance like the bones, but softer, and more elastic.

In order to convey a more clear and correct idea of the form, relative position, and connection, of the bones constituting the human framework, the engraving on page 70, (Fig. 1,) is given.[Pg 70]

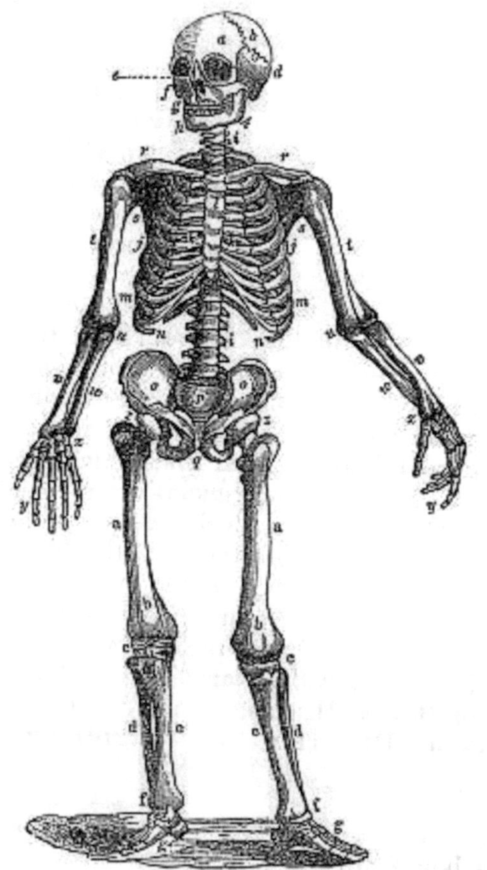

Fig. 1.

By the preceding engraving, it will be seen, that the *cranium*, or *skull*, consists of several distinct pieces, which are united by sutures, (or seams,) as represented by the zigzag lines; *a*, being the *frontal bone*; *b*, the *parietal bone*; *c*, the *temporal bone*; and *d*, the place of the *occipital bone*, which forms the back part of the head, and therefore is not seen in the engraving. The *nasal bones*, or bones of the nose, are shown at *e*; *f*, is the *cheek bone*; *g*, the *upper*, and *h*, the *lower, jaw bones*; *i, i*, the *spinal column*, or back bone, consisting of numerous[Pg 71] small bones, called *vertebræ*; *j, j*, the seven *true ribs*, which are fastened to the spine, behind, and by the *cartilages, k, k*, to the *sternum*, or *breast bone, l*, in front; *m, m*, are the first three *false ribs*, which are so called, because they are not united directly to the breast bone, but by cartilages to the seventh true rib; *n, n*, are the lower two *false*, which are also called *floating, ribs*, because they are not connected with the breast bone, nor the other ribs, in front; *o, o, p, q*, are the bones of the *pelvis*, which is the foundation on which the spine rests; *r, r*, are the *collar bones; s, s*, the *shoulder blades; t, t*, the bones of the *upper arm; u, u*, the *elbow joints*, where the bones of the upper arm and fore arm are united in such a way that they can move like a hinge; *vw, vw*, are the bones of the *fore arm; x, x*, those of the *wrists; y, y*, those of the *fingers; z, z*, are the round heads of the thigh bones, where they are inserted into the sockets of the bones of the pelvis, giving motion in every direction, and forming the *hip joint*; a b, a b, are the *thigh bones*; c, c, the *knee joints*; d e, d e, the *leg bones*; f, f, the *ankle joints*; g, g, the *bones of the foot*.

The bones are composed of two substances, — one animal, and the other mineral. The animal part is a very fine network, called the *cellular membrane*. In this, are deposited the harder mineral substances, which are composed principally of carbonate and phosphate of lime. In very early life, the bones consist chiefly of the animal part, and are then soft and pliant. As the child advances in age, the bones grow harder, by the gradual deposition of the phosphate of lime, which is supplied by the food, and carried to the bones by the blood. In old age, the hardest material preponderates; making the bones more brittle than in earlier life.

As we shall soon have occasion to refer, particularly, to the spinal, or vertebral column, and the derangement to which it is liable, we give, on page 72, representations of the different classes of vertebræ; viz. the *cervical*, (from the Latin, *cervix*, the neck,) the *dorsal*, (from *dorsum*, the back,) and *lumbar*, (from *lumbus*, the loins.)[Pg 72]

Fig.

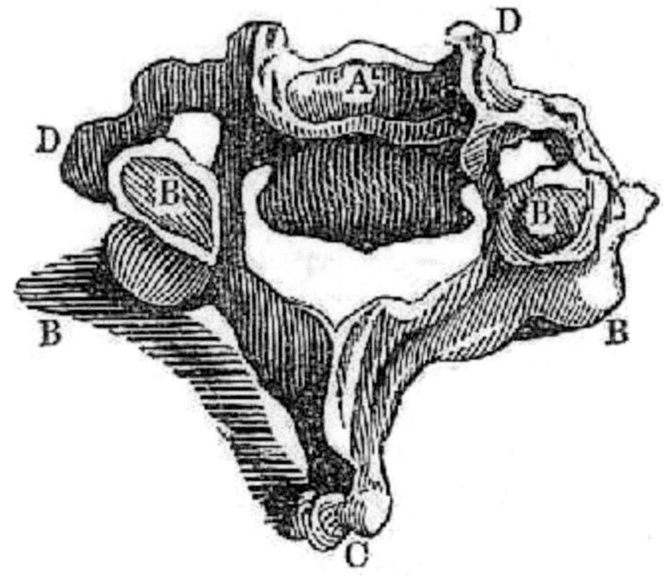

2.

Fig. 2, represents one of the *cervical vertebræ*. Seven of these, placed one above another, constitute that part of the spine which is in the neck.

Fig.

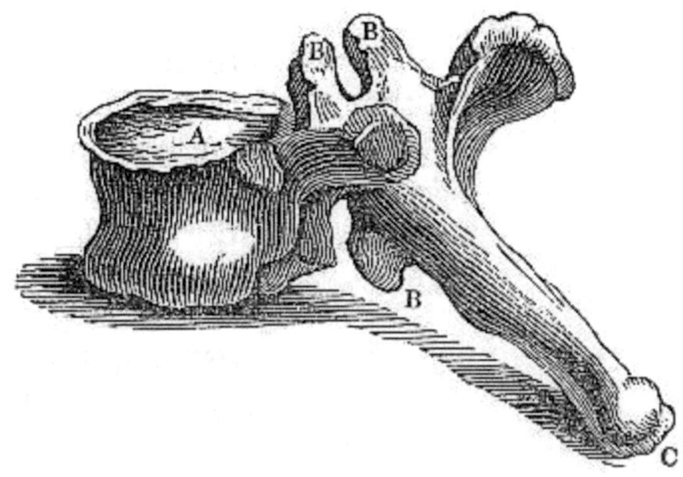

3.

Fig. 3, is one of the *dorsal vertebræ*, twelve of which, form the central part of the spine.

Fig.

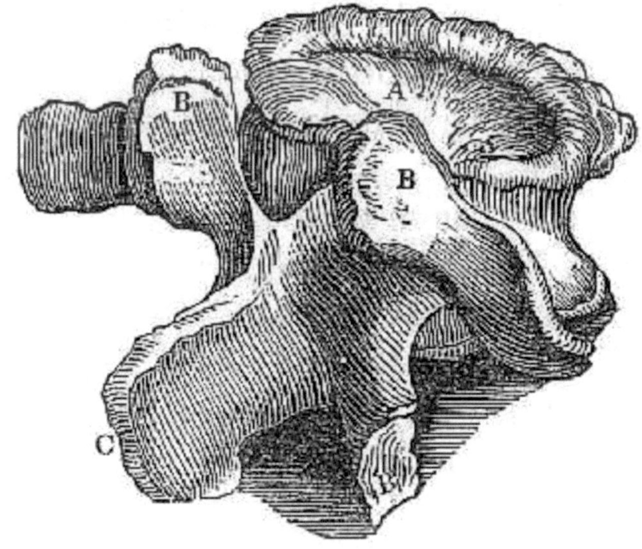

4.

Fig. 4, represents one of the *lumbar vertebræ*, (five in number,) which are immediately above the sacrum. These vertebræ are so fastened, that the spine can bend, in any direction; and the muscles of the trunk are used in holding it erect, or in varying its movements.

By the drawings here presented, it will be seen, that the vertebræ of the neck, back, and loins, differ somewhat in size and shape, although they all possess the same constituent parts; thus, A, in each, represents the body of the vertebræ; B, the articulating processes, by which each is joined to its fellow, above and below it; C, the spinous process, or that part of the vertebræ, which forms the ridge to be felt, on pressure, the whole length of the centre of the back. The back bone receives its name, *spine*, or *spinal column*, from these spinous processes.

It is the universal law of the human frame, that *exercise* is indispensable to the health of the several parts. Thus, if a blood-vessel be tied up, so as not to be used, it shrinks, and becomes a useless string; if a muscle be condemned to inaction, it shrinks in size, and diminishes in power; and thus it is also with the bones. Inactivity produces softness, debility, and unfitness for the functions they are designed to perform. This is one of the causes of the curvature of the spine, that common and pernicious defect in the females of America. From inactivity, the bones of the spine become soft and yielding; and then, if the person is often placed, for a length of time, in positions that throw the weight of the body unequally on certain portions of the spine, they yield to this frequent compression, and a distortion ensues. The positions taken by young persons, when learning to write or draw, or to play on the guitar, harp, or piano, and the position of the body when sleeping on one side, on high pillows, all tend to produce this effect, by throwing the weight of the body unequally, and for a length of time, on particular parts of the spine.[Pg 74]

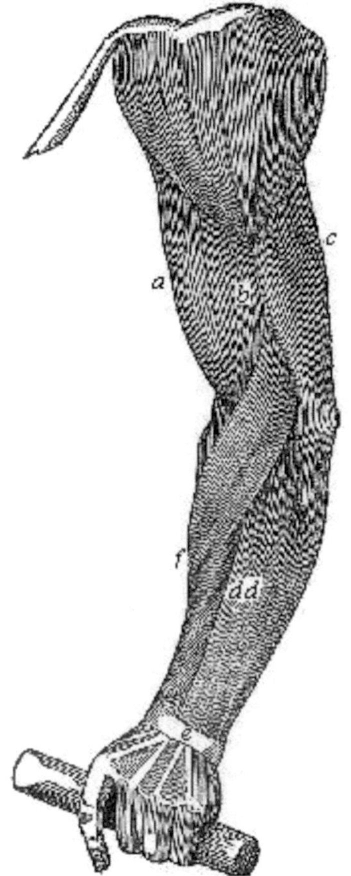

Fig. 5.

MUSCLES.

The muscles are the chief organs of motion, and consist of collections of fine fibres or strings, united in [Pg 75]casings of membrane or thin skin. They possess an elastic power, like India rubber, which enables them to extend and contract. The red meat in animals consists of muscles. Every muscle has connected with it nerves, veins, and arteries; and those designed to move the bones, are fastened to them by tendons at their extremities. The muscles are laid over each other, and are separated by means of membranes and layers of fat, which enable them to move easily, without interfering with each other.

The figure on page 74, represents the muscles of the arm, as they appear when the skin and fat are removed. The muscles *a* and *b* are attached, at their upper ends, to the bone of the arm, and by their lower ends to the upper part of the fore arm, near the elbow joint. When the fibres of these muscles contract, the middle part of them grows larger, and the arm is bent at the elbow. The muscle *c*, is, in like manner, fastened, by its upper end, to the shoulder blade and the upper part of the arm, and by its lower end to one of the bones of the fore arm, near the elbow. When the arm is bent, and we wish to straighten it, it is done by contracting this muscle. The muscles *d*, *d*, are fastened at one end near the elbow joint, and at the other near the ends of the fingers; and on the back of the hand are reduced in size, appearing like strong cords. These cords are called *tendons*. They are employed in straightening the fingers, when the hand is shut. These tendons are confined by the ligament or band, *e*, which binds them down, around the wrist, and thus enables them to act more efficiently, and secures beauty of form to the limb. The muscles at *f*, are those which enable us to turn the hand and arm outward. Every different motion of the arm has one muscle to produce it, and another to restore the limb to its natural position. Those muscles which bend the body are called *flexors*; those which straighten it, *extensors*. When the arm is thrown up, one set of muscles is used; to pull it down, another set: when it is thrown forward, a still different set is used; when it [Pg 76]is thrown back, another, different from the former; when the arm turns in its socket, still another set is used; and thus every different motion of the body is made by a different set of muscles. All these muscles are compactly and skilfully arranged, so as to work with perfect ease. Among them, run the arteries, veins, and nerves, which supply each muscle with blood and nervous power, as will be hereafter described. The size and strength of the muscles depend greatly on their frequent exercise. If left inactive, they grow thin and weak, instead of giving the plumpness to the figure, designed by Nature. The delicate and feeble appearance of many American women, is chiefly owing to the little use they make of their muscles. Many a pale, puny, shad-shaped girl, would have become a plump, rosy, well-formed person, if half the exercise, afforded to her brothers in the open air, had been secured to her, during childhood and youth.

NERVES.

The nerves are the organs of sensation. They enable us to see, hear, feel, taste, and smell; and also combine with the bones and muscles in producing motion.

The first engraving, on p. 77, (Fig. 6,) is a vertical section of the skull, and of the spinal column, or back bone, which supports the head, and through which runs the spinal cord, whence most of the nerves originate. It is a side view, and represents the head and spine, as they would appear, if they were cut through the middle, from front to back. Fig. 7, exhibits them as they would appear, if viewed from *behind*. In Fig. 6, *a*, represents the *cerebrum*, or great brain; *b*, the *cerebellum*, or little brain, which is situated directly under the great brain, at the back and lower part of the head; *c, d, e*, is the spinal marrow, which is connected with the brain at *c*, and runs through the whole length of the spinal column. This column consists, as has already been stated, of a large number of small bones, *f, f*, called *vertebræ*, laid one above another, and fastened together by *cartilage*, or *gristle*, *g*, between them.

[Pg 77]

Fig. 6.

Fig. 7.

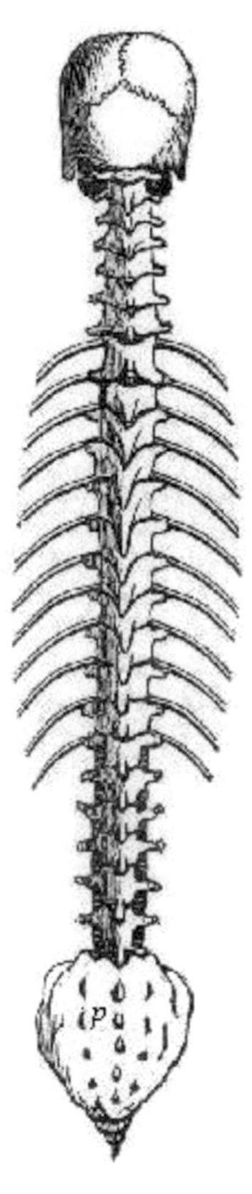

[Pg 78]

Between each two vertebræ, or spinal bones, there issues from the spine, on each side, a pair of nerves. The lower broad part of the spine, (see *p*, Fig. 1, p. 70, and Fig. 7, p. 77,) is called the *sacrum*; in this, are eight holes, through which the lower pairs of nerves pass off.

The nerves of the head and lungs run directly from the brain; those of all other parts of the body proceed from the spine, passing out in the manner already mentioned.

The nerves which thus proceed from the spine, branch out, like the limbs and twigs of a tree, till they extend over the whole body; and, so minutely are they divided and arranged, that a point, destitute of a nerve, cannot be found on the skin.

Some idea of the ramifications of the nerves, may be obtained by reference to the following engraving, (Fig. 8.) In this, A, A, represents the *cerebrum*, or great brain; B, B, the *cerebellum*, or little brain; (see also *a, b*, in Fig. 6;) C, C, represents the union of the fibres of the cerebrum; D, D, the union of the two sides of the cerebellum; E, E, E, the spinal marrow, which passes through the centre of the spine, (as seen at *c, d, e*, in Fig. 6;) 1, 2, 3, 4, 5, 6, branches of the nerves going to different parts of the body. As the nerves are the organs of sensation, all *pain* is an affection of some portion of the nerves. The health of the nerves depends very greatly on the exercise of the muscles, with which they are so intimately connected. This shows the reason why the *headache, tic douloureux*, diseases of the *spine*, and other nervous affections, are so common among American women. Their inactive habits, engender a debility of the nervous system, and these diseases follow, as the consequence.[Pg 79]

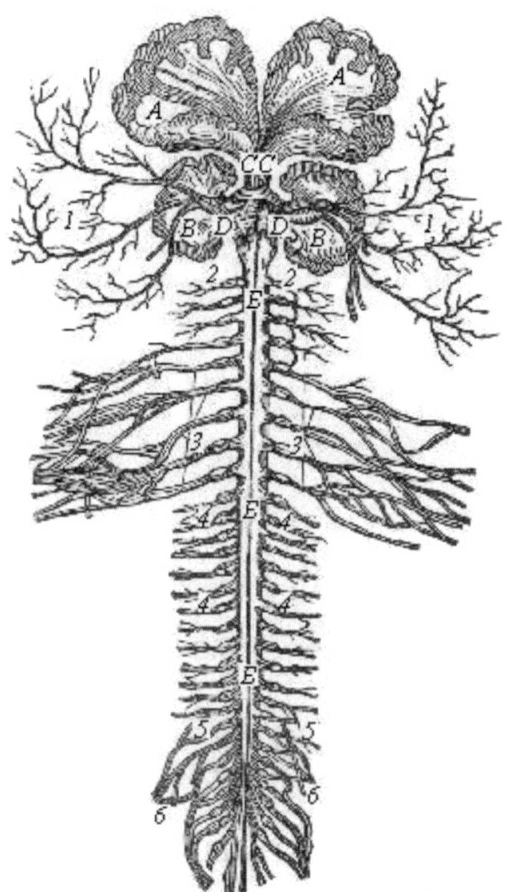

Fig. 8.

[Pg 80] It can be seen, by a reference to the side view, represented on page 77, (Fig. 6,) that the spine is naturally curved back and forward. When, from want of exercise, its bones are softened, and the muscles weakened, the spine acquires an improper curve, and the person becomes what is called *crooked*, having the neck projected forward, and, in some cases, having the back convex, where it should be concave. Probably one half of the American women have the head thus projecting forward, instead of carrying it in the natural, erect position, which is both graceful and dignified.

The curvature of the spine, spoken of in this work as so common, and as the cause of so many diseases among American women, is what is denominated the *lateral curvature*, and is much more dangerous than the other distortion. The indications of this evil, are, the projection of one shoulder blade more than the other, and, in bad cases, one shoulder being higher, and the hip on the opposite side more projecting, than the other. In this case, the spine, when viewed from behind, instead of running in a straight line, (as in Fig. 7 and 9,) is curved somewhat, as may be seen in Figures 10 and 11.

This effect is occasioned by the softness of the bones, induced by want of exercise, together with tight dressing, which tends to weaken the muscles that are thus thrown out of use. Improper and long continued positions in drawing, writing, and sleeping, which throw the weight of the body on one part of the spine, induce the same evil. This distortion is usually accompanied with some consequent disease of the nervous system, or some disarrangement of the internal organs.

By comparing Figures 9 and 11, the difference between a natural and distorted spine will be readily perceived. In Fig. 10, the curved line shows the course of the spine, occasioned by distortion; the perpendicular line, in this and Fig. 11, indicates the true direction of the spine; the horizontal lines show that one shoulder and hip are forced from their proper level.[Pg 81]

Fig. 9.

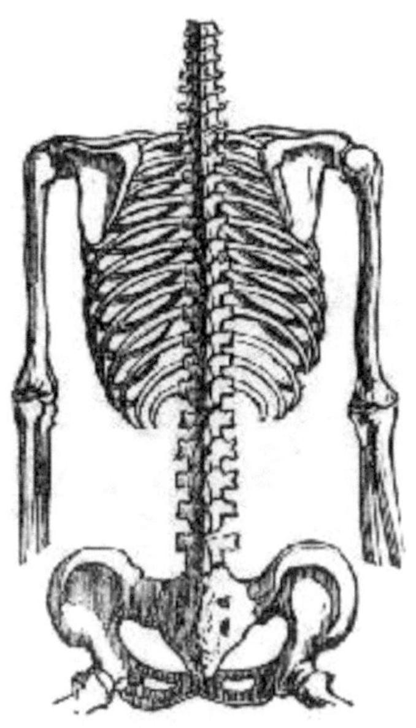

Fig. 10.

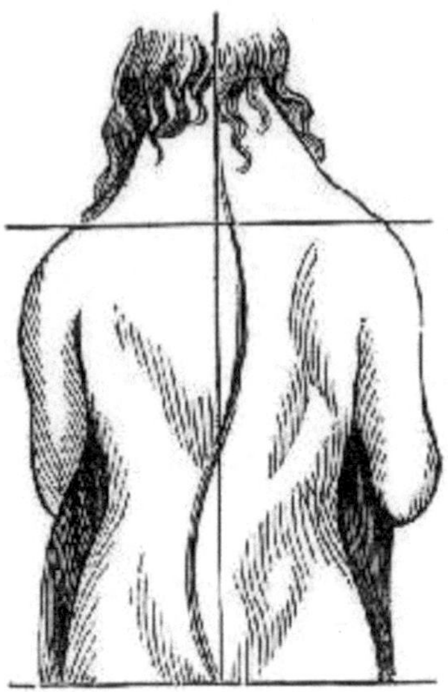

Fig. 11.

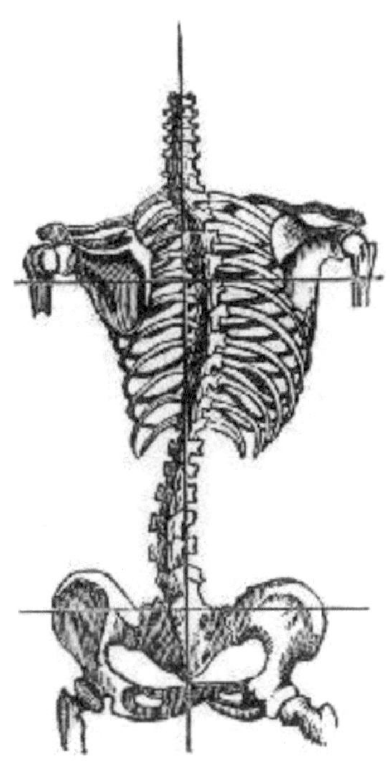

BLOOD-VESSELS.

The blood is the fluid into which our food is changed, and which is employed to minister nourishment to the whole body. For this purpose, it is carried to every part of the body, by the arteries; and, after it has [Pg 82]given out its nourishment, returns to the heart, through the veins.

The subjoined engraving, (Fig. 12,) which presents a rude outline of the vascular system, will more clearly illustrate this operation, as we shall presently show.

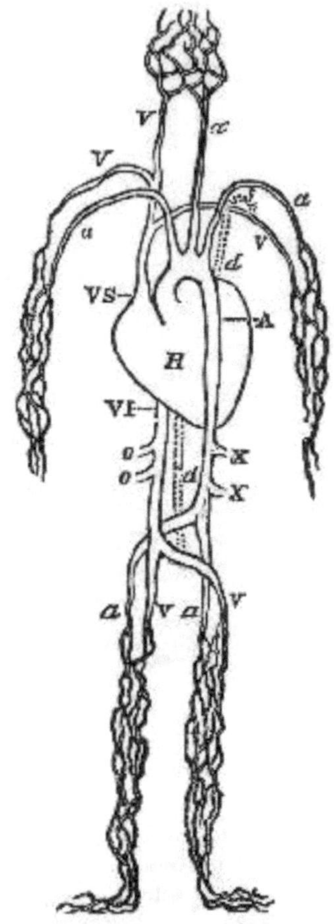

Fig. 12.

[Pg 83]

Before entering the heart, the blood receives a fresh supply of nourishment, by a duct which leads from the stomach. The arteries have their origin from the heart, in a great trunk, called the *aorta*, which is the parent of all the arteries, as the spinal marrow is the parent of the nerves which it sends out. When the arteries have branched out into myriads of minute vessels, the blood which is in them passes into as minute veins; and these run into each other, like the rills and branches of a river, until they are all united in two great veins, which run into the heart. One of these large receivers, called the *vena cava superior*, or *upper vena cava*, brings back the blood from the arms and head, the other, the *vena cava inferior*, or *lower vena cava*, brings back the blood from the body and lower limbs.

In the preceding figure, H, is the heart, which is divided into four compartments; two, called *auricles*, used for receiving the blood, and two, called *ventricles*, used for sending out the blood. A, is the *aorta*, or great artery, which sends its branches to every part of the body. In the upper part, at *a, a, a*, are the main branches of the *aorta*, which go to the head and arms. Below, at *a, a*, are the branches which go to the lower limbs. The branches which set off at X, X, are those by which the intestines are supplied by vessels from the *aorta*. Every muscle in the whole body, all the organs of the body, and the skin, are supplied by branches sent off from this great *artery*. When the blood is thus dispersed through any organ, in minute vessels, it is received, at their terminations, by numerous minute veins, which gradually unite, forming larger branches, till they all meet in either the upper or lower *vena cava*, which returns the blood to the heart. V I, is the *vena cava inferior*, which receives the blood from the veins of the lower parts of the body, as seen at v, v. The blood, sent into the lower limbs from the *aorta*, is received by minute veins, which finally unite at v, v, and thus it is emptied through the lower *vena cava* into the heart: *o, o*, represent the points of entrance of those tributaries of the *vena cava*, [Pg 84]which receive that blood from the intestines, which is sent out by the *aorta* at X, X. In the upper part, V S, is the *vena cava superior*, which receives the blood from the head and arms; v, v, v, are the tributaries of the upper *vena cava*, which bring the blood back from the head and arms; *d, d*, represents the course of the *thoracic duct*, a delicate tube by which the chyle is carried into the blood, as mentioned on page 89; *t*, shows the place where this duct empties into a branch of the *vena cava*.

It thus appears, that wherever a branch of the *aorta* goes to carry blood, there will be found a tributary of the upper or lower *vena cava*, to bring it back.

The succeeding engravings, will enable the reader to form a more definite idea of this important function of the system,—the circulation of the blood. The heart, in man, and in all warm-blooded animals, is double, having two auricles and two ventricles. In animals with cold blood, (as fishes,) the heart is single, having but one auricle and one ventricle. Fig. 13, represents the double heart as it appears when the two sides are separated, and also the great blood-vessels; those on the left of the figure being on the right side of the body, and *vice versa*. The direction of the blood is represented by the arrows. A, represents the *lower vena cava*, returning the blood from the lower parts of the body, and L, the *upper vena cava*, returning the blood from the head and arms. B, is the *right sinus*, or *auricle*, into which the returned blood is poured. From this cavity of the heart, the blood is carried into the *right ventricle*, C; and from this ventricle, the *pulmonary arteries*, D, convey into the lungs the blood which is returned from the body. These five vessels, A, B, C, D, and L, belong to the right side of the heart, and contain the venous or dark-colored blood, which has been through the circulation, and is now unfit for the uses of the system, till it has passed through the lungs.[Pg 85]

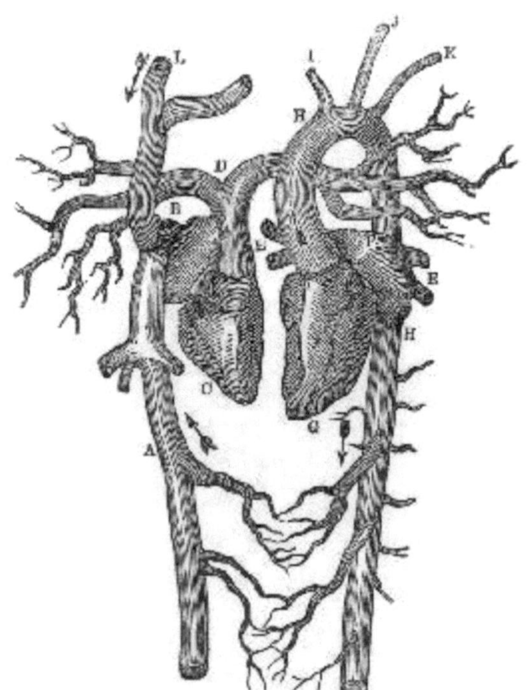

Fig. 13.

94

When the blood reaches the lungs, and is exposed to the action of the air which we breathe, it throws off its impurities, becomes bright in color, and is then called arterial blood. It then returns to the left side of the heart, (on the right of the engraving,) by the pulmonary veins E, E, (also seen at *m, m*, Fig. 15,) into the left auricle F, whence it is forced into the ventricle, G. From the left ventricle, proceeds the *aorta*, H, H, which is the great artery of the body, and conveys the blood to every part of the system. I, J, K, are branches of the aorta, going to the head and arms.[Pg 86]

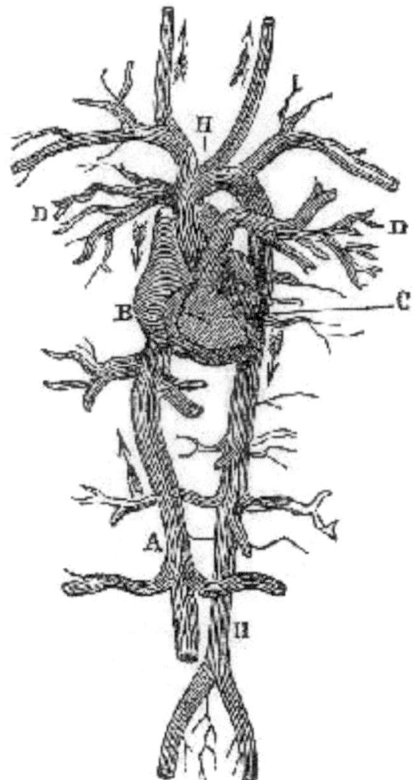

Fig. 14.

Fig. 14, represents the heart, with its two sides united as in nature; and will be understood from the description of Fig. 13.

On the opposite page, Fig. 15, represents the heart, with the great blood-vessels, on a still larger scale; *a*, being the *left ventricle*; *b*, the *right ventricle*; *c, e, f*, the *aorta*, or great artery, rising out of the left ventricle; *g, h, i*, the branches of the aorta, going to the head and arms; *k, l, l*, the *pulmonary artery*, and its branches; *m, m, veins of the lungs*, which bring the blood back from the lungs to the heart; *n*, *right auricle*; *o, vena cava inferior*; *p*, veins returning blood from the liver and bowels; *q*, the *vena cava superior*; *r*, the *left auricle*; *s*, the left *coronary artery*, which distributes the blood exclusively to the substance of the heart.[Pg 87]

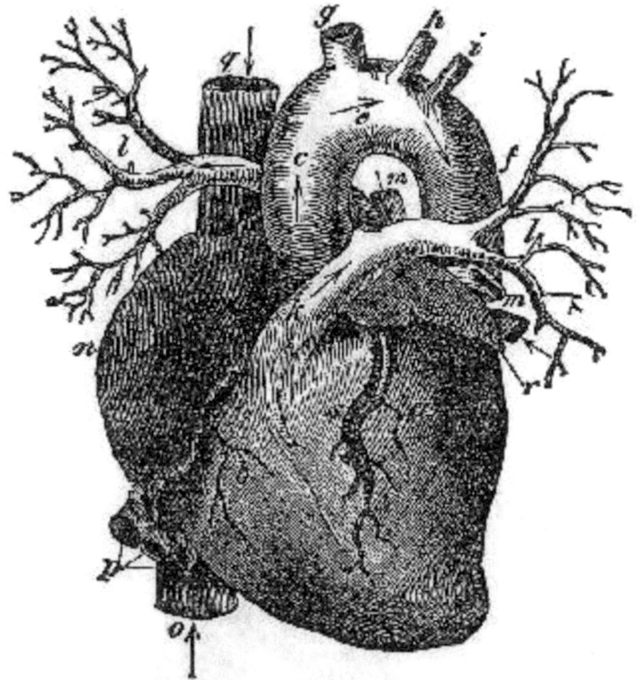

Fig. 15.

ORGANS OF DIGESTION AND RESPIRATION.

Digestion and respiration are the processes, by which the food is converted into blood for the nourishment of the body. The engraving on the next page (Fig. 16) shows the organs by which these operations are performed.

In the lower part of the engraving, is the stomach, marked S, which receives the food through the *gullet*, marked G. The latter, though in the engraving it is cut off at G, in reality continues upwards to the throat. The stomach is a bag composed of muscles, nerves, and blood-vessels, united by a material similar to that which forms the skin. As soon as food enters the stomach, its nerves are excited to perform their proper function of stimulating the muscles. A muscular (called the *peristaltic*) motion immediately commences, by which the stomach propels its contents around the whole of its circumference, once in every three minutes.[Pg 88]

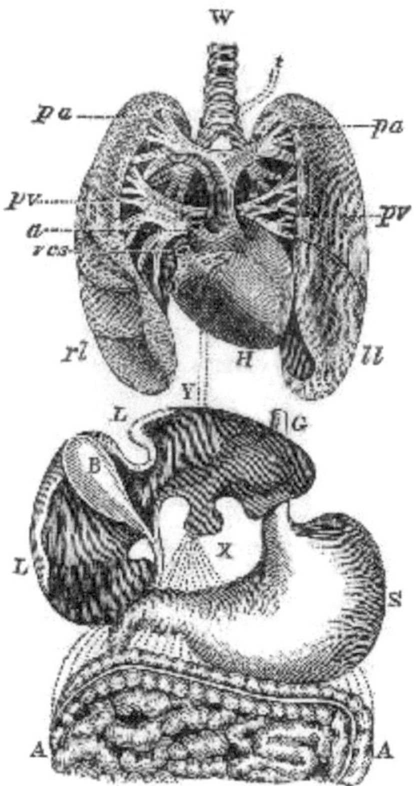

Fig. 16.

This movement of the muscles attracts the blood from other parts of the system; for the blood always hastens to administer its supplies to any organ which is called to work. The blood-vessels of the stomach are soon distended with blood, from which the *gastric juice* is secreted by minute vessels in the coat of the stomach. [Pg 89]This mixes with the food, and reduces it to a soft pulpy mass, called chyme. It then passes through the lower end of the stomach, into the intestines, which are folded up in the abdomen, and the upper portion, only, of which, is shown in the engraving, at A, A. The organ marked L, L, is the liver, which, as the blood passes through its many vessels, secretes a substance called *bile*, which accumulates in the gall-bladder, marked B. After the food passes out of the stomach, it receives from the liver a portion of bile, and from the *pancreas* the *pancreatic juice*. The pancreas does not appear in this drawing, being concealed behind the stomach. These two liquids separate the substance which has passed from the stomach, into two different portions. One is a light liquid, very much like cream in appearance, and called *chyle*, of which the blood is formed; the other is a more solid substance, which contains the refuse and useless matter, with a smaller portion of nourishment; and this, after being further separated from the nourishing matter which it contains, is thrown out of the body. There are multitudes of small vessels, called *lacteals*, which, as these two mixed substances pass through the long and winding folds of the intestines in the abdomen, absorb the chyle, and convey it to the *thoracic duct*, which runs up close by the spine, and carries the chyle, thus received, into a branch of the *vena cava superior*, at *t*, whence it is mingled with the blood going into the heart. In this engraving, the *lacteals* and *thoracic duct* are not shown; but their position is indicated by the dotted lines, marked X, Y; X, being the lacteals, and Y, the thoracic duct.

In the upper half of the engraving, H represents the heart; *a*, the commencement of the *aorta*; *v c s*, the termination of the *vena cava superior*. On each side of the heart, are the lungs; *l l*, being the left lobe, and *r l*, the right lobe. They are composed of a network of air-vessels, blood-vessels, and nerves. W, represents the *trachea*, or *windpipe*, through which, the air we breathe is conducted to the lungs. It branches out [Pg 90]into myriads of minute vessels, which are thus filled with air every time we breathe. From the heart, run the *pulmonary arteries*, marked *p a*. These enter the lungs and spread out along-side of the branches of the air-vessels, so that every air-vessel has a small artery running side by side with it. When the two *vena cavas* empty the blood into the heart, the latter contracts, and sends this blood, through these pulmonary arteries, into the lungs.

As the air and blood meander, side by side, through the lungs, the superabundant carbon and hydrogen of the blood combine with the oxygen of the air, forming carbonic acid gas, and water, which are thrown out of the lungs at every expiration. This is the process by which the chyle is converted into arterial blood, and the venous blood purified of its excess of carbon and hydrogen. When the blood is thus prepared, in the lungs, for its duties, it is received by the small *pulmonary veins*, which gradually unite, and bring the blood back to the heart, through the large *pulmonary veins*, marked *p v, p v*.

On receiving this purified blood from the lungs, the heart contracts, and sends it out again, through the *aorta*, to all parts of the body. It then makes another circuit through every part, ministering to the wants of all, and is afterwards again brought back by the veins to receive the fresh chyle from the stomach, and to be purified by the lungs.

The throbbing of the heart is caused by its alternate expansion and contraction, as it receives and expels the blood. With one throb, the blood is sent from the right ventricle into the lungs, and from the left ventricle into the aorta.

Every time we inspire air, the process of purifying the blood is going on; and every time we expire the air, we throw out the redundant carbon and hydrogen, taken from a portion of the blood. If the waist is compressed by tight clothing, a portion of the lungs be compressed, so that the air-vessels cannot be filled. This [Pg 91]prevents the perfect purification and preparation of the blood, so that a part returns back to the heart unfitted for its duties. This is a slow, but sure, method, by which the constitution of many a young lady is so undermined that she becomes an early victim to disease and to the decay of beauty and strength. The want of *pure air* is another cause, of the debility of the female constitution. When air has been rendered impure, by the breath of several persons, or by close confinement, it does not purify the blood properly. Sleeping in close chambers, and sitting in crowded and unventilated schoolrooms, are frequent causes of debility in the constitution of young persons.

OF THE SKIN.

The skin is the covering of the body, and has very important functions to perform. It is more abundantly supplied with nerves and blood-vessels than any other part; and there is no spot of the skin where the point of the finest needle would not pierce a nerve and blood-vessel. Indeed, it may be considered as composed chiefly of an interlacing of minute nerves and blood-vessels, so that it is supposed there is more nervous matter in the skin, than in all the rest of the body united, and that the greater portion of the blood flows through the skin.

The whole animal system is in a state of continual change and renovation. Food is constantly taken into the stomach, only a portion of which is fitted for the supply of the blood. All the rest has to be thrown out of the system, by various organs designed for this purpose. These organs are, — the lungs, which throw off a portion of useless matter when the blood is purified; the kidneys, which secrete liquids that pass into the bladder, and are thrown out from the body by that organ; and the intestines, which carry off the useless and more solid parts of the food, after the lacteals have drawn off the chyle. In addition to these organs, the skin has a similar duty to perform; and as it has [Pg 92]so much larger a supply of blood, it is the chief organ in relieving the body of the useless and noxious parts of the materials which are taken for food.

Various experiments show, that not less than a pound and four ounces of waste matter is thrown off by the skin every twenty-four hours. This is according to the lowest calculation. Most of those, who have made experiments to ascertain the quantity, represent it as much greater; and all agree, that the skin throws off more redundant matter from the body, than the whole of the other organs together. In the ordinary state of the skin, even when there is no apparent perspiration, it is constantly exhaling waste matter, in a form which is called *insensible perspiration*, because it cannot be perceived by the senses. A very cool mirror, brought suddenly near to the skin, will be covered, in that part, with a moisture, which is this effluvium thus made visible. When heat or exercise excites the skin, this perspiration is increased, so as to be apparent to the senses. This shows the reason why it is so important frequently to wash the entire surface of the body. If this be neglected, the pores of the skin are closed by the waste matter thrown from the body, and by small particles of the thin scarfskin, so that it cannot properly perform its duties. In this way, the other organs are made to work harder, in order to perform the labor the skin would otherwise accomplish, and thus the lungs and bowels are often essentially weakened.

Another office of the skin, is, to regulate the heat of the body. The action of the internal organs is constantly generating heat; and the faster the blood circulates, the greater is the heat evolved. The perspiration of the skin serves to reduce and regulate this heat. For, whenever any liquid changes to a vapor, it absorbs heat from whatever is nearest to it. The faster the blood flows, the more perspiration is evolved. This bedews the skin with a liquid, which the heat of the body turns to a vapor; and in this change, that [Pg 93]heat is absorbed. When a fever takes place, this perspiration ceases, and the body is afflicted with heat. Insensible perspiration is most abundant during sleep, after eating, and when friction is applied to the skin. Perspiration is performed by the terminations of minute arteries in every part of the skin, which exude the perspiration from the blood.

The skin also performs another function. It is provided with a set of small vessels, called *absorbents*, which are exceedingly abundant and minute. When particular substances are brought in contact with the skin, these absorbents take up some portions and carry them into the blood. It is owing to this, that opium, applied on the skin, acts in a manner similar to its operation when taken into the stomach. The power of absorption is increased by friction; and this is the reason that liniments are employed, with much rubbing, to bruises and sprains. The substance applied is thus introduced into the injured part, through the absorbents. This shows another reason for frequent washing of the skin, and for the frequent changes of the garment next the skin. Otherwise portions of the noxious matter, thrown out by the skin, are reabsorbed into the blood, and are slow but sure causes of a decay of the strength of the system.

The skin is also provided with small follicles, or bags, which are filled with an oily substance. This, by gradually exuding over the skin, prevents water from penetrating and injuring its texture.

The skin is also the organ of touch. This office is performed through the instrumentality of the nerves of feeling, which are spread over all parts of the skin.

This general outline of the construction of the human frame is given, with reference to the practical application of this knowledge in the various cases where a woman will be called upon to exercise her own unaided judgement. The application will be further pointed out, in the chapters on Food, Dress, Cleanliness, Care of the Sick, and Care of Infants.

[Pg 94]

CHAPTER VI.
ON HEALTHFUL FOOD.

The person who decides what shall be the food and drink of a family, and the modes of preparation, is the one who decides, to a greater or less extent, what shall be the health of that family. It is the opinion of most medical men, that intemperance in eating is the most fruitful of all causes of disease and death. If this be so, the woman who wisely adapts the food and cooking of her family to the laws of health, removes the greatest risk which threatens the lives of those under her care.

To exhibit this subject clearly, it will be needful to refer, more minutely, to the organization and operation of the digestive organs.

It is found, by experiment, that the supply of gastric juice, furnished from the blood, by the arteries of the stomach, is proportioned, not to the amount of food put into the stomach, but to the wants of the body; so that it is possible to put much more into the stomach than can be digested. To guide and regulate in this matter, the sensation called *hunger* is provided. In a healthy state of the body, as soon as the blood has lost its nutritive supplies, the craving of hunger is felt, and then, if the food is suitable, and is taken in the proper manner, this sensation ceases, as soon as the stomach has received enough to supply the wants of the system. But our benevolent Creator, in this, as in our other duties, has connected enjoyment with the operation needful to sustain our bodies. In addition to the allaying of hunger, the gratification of the palate is secured, by the immense variety of food, some articles of which are far more agreeable than others.

This arrangement of Providence, designed for our happiness, has become, either through ignorance, or [Pg 95]want of self-control, the chief cause of the various diseases and sufferings, which afflict those classes who have the means of seeking a variety to gratify the palate. If mankind had only one article of food, and only water to drink, though they would have less enjoyment in eating, they would never be tempted to put any more into the stomach, than the calls of hunger required. But the customs of society, which present an incessant change, and a great variety of food, with those various condiments which stimulate appetite, lead almost every person very frequently to eat merely to gratify the palate, after the stomach has been abundantly supplied, so that hunger has ceased.

When too great a supply of food is put into the stomach, the gastric juice dissolves only that portion which the wants of the system demand. The remainder is ejected, in an unprepared state; the absorbents take portions of it into the system; and all the various functions of the body, which depend on the ministries of the blood, are thus gradually and imperceptibly injured. Very often, intemperance in eating produces immediate results, such as colic, headaches, pains of indigestion, and vertigo. But the more general result, is, a gradual undermining of all parts of the human frame; thus imperceptibly shortening life, by so weakening the constitution, that it is ready to yield, at every point, to any uncommon risk or exposure. Thousands and thousands are passing out of the world, from diseases occasioned by exposures, which a healthy constitution could meet without any danger. It is owing to these considerations, that it becomes the duty of every woman, who has the responsibility of providing food for a family, to avoid a variety of tempting dishes. It is a much safer rule, to have only one kind of healthy food, for each meal, than the abundant variety which is usually met at the tables of almost all classes in this Country. When there is to be any variety of dishes, they ought not to be successive, but so arranged, as to give the opportunity of selection. How [Pg 96]often is it the case, that persons, by the appearance of a favorite article, are tempted to eat, merely to gratify the palate, when the stomach is already adequately supplied. All such intemperance wears on the constitution, and shortens life. It not unfrequently happens, that excess in eating produces a morbid appetite, which must constantly be denied.

But the organization of the digestive organs demands, not only that food be taken in proper quantities, but that it be taken at proper times.

It has before been shown, that, as soon as the food enters the stomach, the muscles are excited by the nerves, and the *peristaltic motion* commences. This is a powerful and constant exercise of the muscles of the stomach, which continues until the process of digestion is complete. During this time, the blood is withdrawn from other parts of the system, to supply the demands of the stomach, which is laboring hard with all its muscles. When this motion ceases, and the digested food has gradually passed out of the stomach, Nature requires that it should have a period of repose. And if another meal be eaten, immediately after one is digested, the stomach is set to work again, before it has had time to rest, and before a sufficient supply of gastric juice is provided.

The general rule, then, is, that three hours be given to the stomach for labor, and two for rest; and in obedience to this, five hours, at least, ought to elapse between every two regular meals. In cases where exercise produces a flow of perspiration, more food is needed to supply the loss; and strong laboring men may safely eat as often as they feel the want of food. So, young and healthy children, who gambol and exercise much, and whose bodies grow fast, may have a more frequent supply of food. But, as a general rule, meals should be five hours apart, and eating between meals avoided. There is nothing more unsafe, and wearing to the constitution, than a habit of eating at any time, merely to gratify the palate. When a tempting [Pg 97]article is presented, every person should exercise sufficient self-denial, to wait till the proper time for eating arrives. Children, as well as grown persons, are often injured, by eating between their regular meals, thus weakening the stomach, by not affording it any time for rest.

In deciding as to *quantity* of food, there is one great difficulty to be met by a large portion of the community. It has been shown, that the exercise of every part of the body is indispensable to its health and perfection. The bones, the muscles, the nerves, the organs of digestion and respiration, and the skin, all demand exercise, in order properly to perform their functions. When the muscles of the body are called into action, all the blood-vessels entwined among them are frequently compressed. As the arteries are so contrived, that the blood cannot run back, this compression hastens it forward, through the veins, towards that organ. The heart is immediately put in quicker motion, to send it into the lungs; and they, also, are thus stimulated to more rapid action, which is the cause of that panting which active exercise always occasions. The blood thus courses with greater celerity through the body, and sooner loses its nourishing properties. Then the stomach issues its mandate of hunger, and a new supply of food must be furnished. Thus it appears, as a general rule, that the quantity of food, actually needed by the body, depends on the amount of muscular exercise taken. A laboring man, in the open fields, probably throws off from his skin ten times the amount of perspirable matter, which is evolved from the skin of a person of sedentary pursuits. In consequence of this, he demands a far greater amount of food and drink.

Those persons, who keep their bodies in a state of health, by sufficient exercise, can always be guided by the calls of hunger. They can eat when they feel hungry, and stop when hunger ceases; and then they will calculate exactly right. But the difficulty is, [Pg 98]that a large part of the community, especially women, are so inactive in their habits, that they seldom feel the calls of hunger. They habitually eat, merely to gratify the palate. This produces such a state of the system, that they have lost the guide which Nature has provided. They are not called to eat, by hunger, nor admonished, by its cessation, when to stop. In consequence of this, such persons eat what pleases the palate, till they feel no more inclination for the article. It is probable, that three fourths of the women, in the wealthier circles, sit down to each meal without any feeling of hunger, and eat merely on account of the gratification thus afforded them. Such persons find their appetite to depend almost solely upon the kind of food on the table. This is not the case with those, who take the exercise which Nature demands. They approach their meals in such a state that almost any kind of food is acceptable.

The question then arises, how are persons, who have lost the guide which Nature has provided, to determine as to the proper amount of food they shall take?

The only rules they can adopt, are of a general nature; founded on the principles already developed. They should endeavor to proportion their food to the amount of the exercise they ordinarily take. If they take but little exercise, they should eat but little food in comparison with those who are much in the open air and take much exercise; and their food should be chiefly vegetable, and not animal. But how often is it seen, that a student, or a man who sits all day in an office, or a lady who spends the day in her parlor and chamber, will sit down to a loaded table, and, by continuing to partake of the tempting varieties, in the end load the stomach with a supply, which a stout farmer could scarcely digest.

But the health of a family depends, not merely on the *quantity* of food taken; but very much, also, on the *quality*. Some kinds of food are very pernicious [Pg 99]in their nature, and some healthful articles are rendered very injurious by the mode of cooking. Persons who have a strong constitution, and take much exercise, may eat almost any thing, with apparent impunity; but young children, who are forming their constitutions, and persons who are delicate, and who take but little exercise, are very dependent for health, on a proper selection of food.

There are some general principles, which may aid in regulating the judgement on this subject.

It is found, that there are some kinds of food which afford nutriment to the blood, and do not produce any other effect on the system. There are other kinds, which are not only nourishing, but *stimulating*, so that they quicken the functions of the organs on which they operate. The condiments used in cookery, such as pepper, mustard, and spices, are of this nature. There are certain states of the system, when these stimulants are beneficial; but it is only in cases where there is some debility. Such cases can only be pointed out by medical men. But persons in perfect health, and especially young children, never receive any benefit from such kind of food; and just in proportion as condiments operate to quicken the labors of the internal organs, they tend to wear down their powers. A person who thus keeps the body working under an unnatural excitement, *lives faster* than Nature designed, and the sooner the constitution is worn out. A woman, therefore, should provide dishes for her family, which are free from these stimulating condiments, and as much as possible prevent their use. It is also found, by experience, that animal food is more stimulating than vegetable. This is the reason why, in cases of fevers, or inflammations, medical men forbid the use of meat and butter. Animal food supplies chyle much more abundantly than vegetable food does; and this chyle is more stimulating in its nature. Of course, a person who lives chiefly on animal food, is under a higher degree of stimulus than if his food was chiefly [Pg 100]composed of vegetable substances. His blood will flow faster, and all the functions of his body will be quickened.

This makes it important to secure a proper proportion of animal and vegetable diet. Some medical men suppose, that an exclusively vegetable diet is proved, by the experience of many individuals, to be fully sufficient to nourish the body; and bring, as evidence, the fact, that some of the strongest and most robust men in the world, are those, who are trained, from infancy, exclusively on vegetable food. From this, they infer, that life will be shortened, just in proportion as the diet is changed to more stimulating articles; and that, all other things being equal, children will have a better chance of health and long life, if they are brought up solely on vegetable food.

But, though this is not the common opinion of medical men, they all agree, that, in America, far too large a portion of the diet consists of animal food. As a nation, the Americans are proverbial for the gross and luxurious diet with which they load their tables; and there can be no doubt that the general health of the Nation would be increased, by a change in our customs in this respect. To take meat but once a day, and this in small quantities, compared with the common practice, is a rule, the observance of which would probably greatly reduce the amount of fevers, eruptions, headaches, bilious attacks, and the many other ailments which are produced or aggravated by too gross a diet.

The celebrated Roman physician, Baglivi, (who, from practising extensively among Roman Catholics, had ample opportunities to observe,) mentions, that, in Italy, an unusual number of people recover their health in the forty days of Lent, in consequence of the lower diet which is required as a religious duty. An American physician remarks, "For every reeling drunkard that disgraces our Country, it contains one hundred gluttons;—persons, I mean, who eat to excess, and suffer in consequence." Another distinguished physician [Pg 101]says, "I believe that every stomach, not actually impaired by organic disease, will perform its functions, if it receives reasonable attention; and when we perceive the manner in which diet is generally conducted, both in regard to *quantity* and *variety* of articles of food and drink, which are mixed up in one heterogeneous mass,—instead of being astonished at the prevalence of indigestion, our wonder must rather be, that, in such circumstances, any stomach is capable of digesting at all."

In regard to articles which are the most easily digested, only general rules can be given. Tender meats are digested more readily than those which are tough, or than many kinds of vegetable food. The farinaceous articles, such as rice, flour, corn, potatoes, and the like, are the most nutritious, and most easily digested. The popular notion, that meat is more nourishing than bread, is a great mistake. Good bread contains one third more nourishment than butcher's meat. The meat is more *stimulating*, and for this reason is more readily digested. A perfectly healthy stomach can digest almost any healthful food; but when the digestive powers are weak, every stomach has its peculiarities, and what is good for one, is hurtful to another. In such cases, experiment, alone, can decide, which are the most digestible articles of food. A person, whose food troubles him, must deduct one article after another, till he learns, by experience, which is the best for digestion. Much evil has been done, by assuming that the powers of one stomach are to be made the rule in regulating every other.

The most unhealthful kinds of food, are those, which are made so by bad cooking; such as sour and heavy bread, cakes, pie-crust, and other dishes consisting of fat, mixed and cooked with flour; also rancid butter, and high-seasoned food. The fewer mixtures there are in cooking, the more healthful is the food likely to be.

There is one caution, as to the *mode* of eating, which seems peculiarly needful to Americans. It is indispensable to good digestion, that food be well chewed and taken slowly. It needs to be thoroughly chewed, in [Pg 102]order to prepare it for the action of the gastric juice, which, by the *peristaltic motion*, will be thus brought into universal contact with the minute portions. It has been found, that a solid lump of food requires much more time and labor of the stomach, than divided substances. It has also been found, that, as each bolus, or mouthful, enters the stomach, the latter closes, until the portion received has had some time to move around and combine with the gastric juice; and that the orifice of the stomach resists the entrance of any more, till this is accomplished. But, if the eater persists in swallowing fast, the stomach yields; the food is then poured in more rapidly than the organ can perform its duty of digestion; and evil results are sooner or later developed. This exhibits the folly of those hasty meals, so common to travellers, and to men of business, and shows why children should be taught to eat slowly.

After taking a full meal, it is very important to health, that no great bodily or mental exertion be made, till the labor of the stomach is over. Intense mental effort draws the blood to the head, and muscular exertions draw it to the muscles; and in consequence of this, the stomach loses the supply which it requires when performing its office. When the blood is thus withdrawn, the adequate supply of gastric juice is not afforded, and indigestion is the result. The heaviness which follows a full meal, is the indication which Nature gives of the need of quiet. When the meal is moderate, a sufficient quantity of gastric juice is exuded in an hour, or an hour and a half; after which, labor of body and mind may safely be resumed.

When undigested food remains in the stomach, and is at last thrown out into the bowels, it proves an irritating substance, producing an inflamed state in the lining of the stomach and other organs. The same effect is produced by alcoholic drinks.

It is found, that the stomach has the power of gradually accommodating its digestive powers to the food it habitually receives. Thus, animals, which live on [Pg 103]vegetables, can gradually become accustomed to animal food; and the reverse is equally true. Thus, too, the human stomach can eventually accomplish the digestion of some kinds of food, which, at first, were indigestible.

But any changes of this sort should be gradual; as those which are *sudden*, are trying to the powers of the stomach, by furnishing matter for which its gastric juice is not prepared.

In regard to the nature of the meals prepared, the breakfast should furnish a supply of liquids, because the body has been exhausted by the exhalations of the night, and demands them more than at any other period. It should not be the heartiest meal, because the organs of digestion are weakened by long fasting, and the exhalations. Dinner should be the heartiest meal, because then the powers of digestion are strengthened, by the supplies of the morning meal. Light and amusing employments should occupy mind and body for an hour or more after a full meal.

But little drink should be taken, while eating, as it dilutes the gastric juice which is apportioned to each quantity of food as it enters the stomach. It is better to take drink after the meal is past.

Extremes of heat or cold are injurious to the process of digestion. Taking hot food or drink, habitually, tends to debilitate all the organs thus needlessly excited. In using cold substances, it is found that a certain degree of warmth in the stomach is indispensable to their digestion; so that, when the gastric juice is cooled below this temperature, it ceases to act. Indulging in large quantities of cold drinks, or eating ice-creams, after a meal, tends to reduce the temperature of the stomach, and thus to stop digestion. This shows the folly of those refreshments, in convivial meetings, where the guests are tempted to load the stomach with a variety, such as would require the stomach of a stout farmer to digest, and then to wind up with ice-creams, thus destroying whatever ability might otherwise have existed, to digest the heavy load. The fittest temperature [Pg 104]for drinks, if taken when the food is in the digesting process, is blood heat. Cool drinks, and even ice, can be safely taken at other times, if not in excessive quantity. When the thirst is excessive, or the body weakened by fatigue, or when in a state of perspiration, cold drinks are injurious. When the body is perspiring freely, taking a large quantity of cold drink has often produced instant death.

Fluids taken into the stomach are not subject to the slow process of digestion, but are immediately absorbed and carried into the blood. This is the reason why drink, more speedily than food, restores from exhaustion. The minute vessels of the stomach inhale or absorb its fluids, which are carried into the blood, just as the minute extremities of the arteries open upon the inner surface of the stomach, and there exude the gastric juice from the blood.

When food is chiefly liquid, (soup, for example,) the fluid part is rapidly absorbed. The solid parts remain, to be acted on by the gastric juice. In the case of St. Martin,[F] in fifty minutes after taking soup, the fluids were absorbed, and the remainder was even thicker than is usual after eating solid food. This is the reason why soups are deemed bad for weak stomachs; as this residuum is more difficult of digestion than ordinary food. In recovering from sickness, beef-tea and broths are good, because the system then demands fluids to supply its loss of blood.

Highly-concentrated food, having much nourishment in a small bulk, is not favorable to digestion, because[Pg 105] it cannot be properly acted on by the muscular contractions of the stomach, and is not so minutely divided, as to enable the gastric juice to act properly. This is the reason, why a certain *bulk* of food is needful to good digestion; and why those people, who live on whale oil, and other highly-nourishing food, in cold climates, mix vegetables and even sawdust with it, to make it more acceptable and digestible. So, in civilized lands, bread, potatoes, and vegetables, are mixed with more highly-concentrated nourishment. This explains why coarse bread, of unbolted wheat, so often proves beneficial. Where, from inactive habits, or other causes, the bowels become constipated and sluggish, this kind of food proves the appropriate remedy. One fact on this subject is worthy of notice. Under the administration of William Pitt, for two years or more, there was such a scarcity of wheat, that, to make it hold out longer, Parliament passed a law, that the army should have all their bread made of unbolted flour. The result was, that the health of the soldiers improved so much, as to be a subject of surprise to themselves, the officers, and the physicians. These last came out publicly, and declared, that the soldiers never before were so robust and healthy; and that disease had nearly disappeared from the army. The civic physicians joined and pronounced it the healthiest bread; and, for a time, schools, families, and public institutions, used it almost exclusively. Even the nobility, convinced by these facts, adopted it for their common diet; and the fashion continued a long time after the scarcity ceased, until more luxurious habits resumed their sway. For this reason, also, soups, gellies, and arrow-root, should have bread or crackers mixed with them. We thus see why children should not have cakes and candies allowed them between meals. These are highly-concentrated nourishments, and should be eaten with more bulky and less nourishing substances. The most indigestible of all kinds of food, are fatty and oily substances; especially if heated. [Pg 106]It is on this account, that pie-crust, and articles boiled and fried in fat or butter, are deemed not so healthful as other food.

The following, then, may be put down as the causes of a debilitated constitution, from the misuse of food. Eating *too much*, eating *too often*, eating *too fast*, eating food and condiments that are *too stimulating*, eating food that is *too warm* or *too cold*, eating food that is *highly-concentrated*, without a proper admixture of less nourishing matter, and eating food that is *difficult of digestion.*

FOOTNOTE:

[F] The individual here referred to, — Alexis St. Martin, — was a young Canadian, of eighteen years of age, of a good constitution, and robust health, who, in 1822, was accidentally wounded by the discharge of a musket, which carried away a part of the ribs, lacerated one of the lobes of the lungs, and perforated the stomach, making a large aperture, which never closed; and which enabled Dr. Beaumont, (a surgeon of the American army, stationed at Michilimackinac, under whose care the patient was placed,) to witness all the processes of digestion and other functions of the body, for several years. The published account of the experiments made by Dr. B., is highly interesting and instructive.

CHAPTER VII.
ON HEALTHFUL DRINKS.

Although intemperance in eating is probably the most prolific cause of the diseases of mankind, intemperance in drink has produced more guilt, misery, and crime, than any other one cause. And the responsibilities of a woman, in this particular, are very great; for the habits and liabilities of those under her care, will very much depend on her opinions and practice.

It is a point fully established by experience, that the full developement of the human body, and the vigorous exercise of all its functions, can be secured without the use of stimulating drinks. It is, therefore, perfectly safe, to bring up children never to use them; no hazard being incurred, by such a course.

It is also found, by experience, that there are two evils incurred, by the use of stimulating drinks. The first, is, their positive effect on the human system. Their peculiarity consists in so exciting the nervous system, that all the functions of the body are accelerated, and the fluids are caused to move quicker than at their natural speed. This increased motion of the animal fluids, always produces an agreeable effect on the mind. The intellect is invigorated, the imagination [Pg 107]is excited, the spirits are enlivened; and these effects are so agreeable, that all mankind, after having once experienced them, feel a great desire for their repetition.

But this temporary invigoration of the system, is always followed by a diminution of the powers of the stimulated organs; so that, though in all cases this reaction may not be perceptible, it is invariably the result. It may be set down as the unchangeable rule of physiology, that stimulating drinks (except in cases of disease) deduct from the powers of the constitution, in exactly the proportion in which they operate to produce temporary invigoration.

The second evil, is, the temptation which always attends the use of stimulants. Their effect on the system is so agreeable, and the evils resulting are so imperceptible and distant, that there is a constant tendency to increase such excitement, both in frequency and power. And the more the system is thus reduced in strength, the more craving is the desire for that which imparts a temporary invigoration. This process of increasing debility and increasing craving for the stimulus that removes it, often goes to such an extreme, that the passion is perfectly uncontrollable, and mind and body perish under this baleful habit.

In this Country, there are five forms in which the use of such stimulants is common; namely, *alcoholic drinks, tea, coffee, opium mixtures*, and *tobacco*. These are all alike, in the main peculiarity of imparting that extra stimulus to the system, which tends to exhaust its powers.

Multitudes in this Nation are in the habitual use of some one of these stimulants; and each person defends the indulgence by these arguments:

First, that the desire for stimulants is a natural propensity, implanted in man's nature, as is manifest from the universal tendency to such indulgences, in every nation. From this, it is inferred, that it is an innocent desire, which ought to be gratified, to some extent, and [Pg 108]that the aim should be, to keep it within the limits of temperance, instead of attempting to exterminate a natural propensity.

This is an argument, which, if true, makes it equally proper to use opium, brandy, tea, or tobacco, as stimulating principles, provided they are used temperately. But, if it be granted that perfect health and strength can be gained and secured without these stimulants, and that their peculiar effect is to diminish the power of the system, in exactly the same proportion as they stimulate it, then there is no such thing as a temperate use, unless they are so diluted, as to destroy any stimulating power; and in this form, they are seldom desired.

The other argument for their use, is, that they are among the good things provided by the Creator, for our gratification; that, like all other blessings, they are exposed to abuse and excess; and that we should rather seek to regulate their use, than to banish them entirely.

This argument is based on the assumption, that they are, like healthful foods and drinks, necessary to life and health, and injurious only by excess. But this is not true; for, whenever they are used in any such strength as to be a gratification, they operate, to a greater or less extent, as stimulants; and, to just such extent, they wear out the powers of the constitution; and it is abundantly proved, that they are not, like food and drink, necessary to health. Such articles are designed for medicine, and not for common use. There can be no argument framed to defend the use of one of them, which will not equally defend all. That men have a love for being stimulated, after they have once felt the pleasurable excitement, and that Providence has provided the means for securing it, are arguments as much in favor of alcohol, opium, and tobacco, as of coffee and tea. All that can be said in favor of the last-mentioned favorite beverages, is, that the danger in their use is not so great. Let any one, who defends one kind of stimulating drink, remember, then, that he uses an argument, which, if it be allowed that stimulants are not needed, [Pg 109]and are injurious, will equally defend all kinds; and that all which can be said in defence of tea and coffee, is, that they *may* be used, so weak, as to do no harm, and that they actually have done less harm than some of the other stimulating narcotics.

The writer is of opinion, that tea and coffee are a most extensive cause of much of the nervous debility and suffering endured by American women; and that relinquishing such drinks would save an immense amount of such suffering. But there is little probability that the present generation will make so decided a change in their habits, as to give up these beverages; and the subject is presented rather in reference to forming the habits of children.

It is a fact, that tea and coffee are, at first, seldom or never agreeable to children. It is the mixture of milk, sugar, and water, that reconciles them to a taste, which in this manner gradually becomes agreeable. Now, suppose that those who provide for a family conclude that it is not *their* duty to give up entirely the use of stimulating drinks, may not the case appear different, in regard to teaching their children to love such drinks? Let the matter be regarded thus:—The experiments of physiologists all prove, that stimulants are not needful to health, and that, as the general rule, they tend to debilitate the constitution. Is it right, then, for a parent to tempt a child to drink what is not needful, when there is a probability that it will prove, to some extent, an undermining drain on the constitution? Some constitutions can bear much less excitement than others; and, in every family of children, there is usually one, or more, of delicate organization, and consequently peculiarly exposed to dangers from this source. It is this child who ordinarily becomes the victim to stimulating drinks. The tea and coffee which the parents and the healthier children can use without immediate injury, gradually sap the energies of the feebler child, who proves either an early victim, or a living martyr to all the sufferings that debilitated nerves inflict. Can it be [Pg 110]right, to lead children, where all allow that there is some danger, and where, in many cases, disease and death are met, when another path is known to be perfectly safe?

Of the stimulating drinks in common use, *black tea* is least injurious, because its flavor is so strong, in comparison with its narcotic principle, that one who uses it, is much less liable to excess. Children can be trained to love milk and water sweetened with sugar, so that it will always be a pleasant beverage; or, if there are exceptions to the rule, they will be few. Water is an unfailing resort. Every one loves it, and it is perfectly healthful.

The impression, common in this Country, that *warm drinks*, especially in Winter, are more healthful than cold, is not warranted by any experience, nor by the laws of the physical system. At dinner, cold drinks are universal, and no one deems them injurious. It is only at the other two meals that they are supposed to be hurtful.

There is no doubt that *warm* drinks are healthful, and more agreeable than cold, at certain times and seasons; but it is equally true, that drinks above blood heat are not healthful. If any person should hold a finger in hot water, for a considerable time, twice every day, it would be found that the finger would gradually grow weaker. The frequent application of the stimulus of heat, like all other stimulants, eventually causes debility. If, therefore, a person is in the habit of drinking hot drinks, twice a day, the teeth, throat, and stomach are gradually debilitated. This, most probably, is one of the causes of an early decay of the teeth, which is observed to be much more common among American ladies, than among those in European countries.

It has been stated to the writer, by an intelligent traveller, who had visited Mexico, that it was rare to meet an individual with even a tolerable set of teeth; and that almost every grown person, he met in the street, had merely remnants of teeth. On inquiry into [Pg 111]the customs of the Country, it was found, that it was the universal practice to take their usual beverage at almost the boiling point; and this, doubtless, was the chief cause of the almost entire want of teeth in that Country. In the United States, it cannot be doubted that much evil is done, in this way, by hot drinks. Most tea-drinkers consider tea as ruined, if it stands until it reaches the healthful temperature for drink.

The following extract from Dr. Andrew Combe, presents the opinion of most intelligent medical men on this subject.[G]

"*Water* is a safe drink for all constitutions, provided it be resorted to in obedience to the dictates of natural thirst, only, and not of habit. Unless the desire for it is felt, there is no occasion for its use during a meal."

"The primary effect of all distilled and fermented liquors, is, to *stimulate the nervous system and quicken the circulation*. In infancy and childhood, the circulation is rapid, and easily excited; and the nervous system is strongly acted upon, even by the slightest external impressions. Hence slight causes of irritation readily excite febrile and convulsive disorders. In youth, the natural tendency of the constitution is still to excitement; and consequently, as a general rule, the stimulus of fermented liquors is injurious."

These remarks show, that parents, who find that stimulating drinks are not injurious to themselves, may mistake in inferring, from this, that they will not be injurious to their children.

Dr. Combe continues thus: "In mature age, when digestion is good and the system in full vigor, if the mode of life be not too exhausting, the nervous functions and general circulation are in their best condition, and require no stimulus for their support. The bodily energy is then easily sustained, by nutritious food and a regular regimen, and consequently artificial excitement [Pg 112]only increases the wasting of the natural strength. In old age, when the powers of life begin to fail, moderate stimulus may be used with evident advantage."

It may be asked, in this connection, why the stimulus of animal food is not to be regarded in the same light, as that of stimulating drinks. In reply, a very essential difference may be pointed out. Animal food furnishes nutriment to the organs which it stimulates, but stimulating drinks excite the organs to quickened action, without affording any nourishment.

It has been supposed, by some, that tea and coffee have, at least, a degree of nourishing power. But it is proved, that it is the milk and sugar, and not the main portion of the drink, which imparts the nourishment. Tea has not one particle of nourishing properties; and what little exists in the coffee-berry, is lost by roasting it in the usual mode. All that these articles do, is simply *to stimulate, without nourishing*.

It is very common, especially in schools, for children to form a habit of drinking freely of cold water. This is a debilitating habit, and should be corrected. Very often, chewing a bit of cracker will stop a craving for drink, better than taking water; and when teachers are troubled with very thirsty scholars, they should direct them to this remedy. A person who exercises but little, requires no drink, between meals, for health; and the craving for it is unhealthful. Spices, wines, fermented liquors, and all stimulating condiments, produce unhealthful thirst.

FOOTNOTE:

[G] The writer would here remark, in reference to extracts made from various authors, that, for the sake of abridging, she has often left out parts of a paragraph, but never so as to modify the meaning of the author. Some ideas, not connected with the subject in hand, are omitted, but none are altered.

CHAPTER VIII.
ON CLOTHING.

It appears, by calculations made from bills of mortality, that one quarter of the human race perishes in infancy. This is a fact not in accordance with the analogy of Nature. No such mortality prevails [Pg 113]among the young of animals; it does not appear to be the design of the Creator; and it must be owing to causes which can be removed. Medical men agree in the opinion, that a great portion of this mortality, is owing to mismanagement, in reference to fresh air, food, and clothing.

At birth, the circulation is chiefly in the vessels of the skin; for the liver and stomach, being feeble in action, demand less blood, and it resorts to the surface. If, therefore, an infant be exposed to cold, the blood is driven inward, by the contracting of the blood-vessels in the skin: and, the internal organs being thus over-stimulated, bowel complaints, croup, convulsions, or some other evil, ensues. This shows the sad mistake of parents, who plunge infants in cold water to strengthen their constitution; and teaches, that infants should be washed in warm water, and in a warm room. Some have constitutions strong enough to bear mismanagement in these respects; but many fail in consequence of it.

Hence we see the importance of dressing infants warmly, and protecting them from exposure to a cold temperature. It is for this purpose, that mothers, now, very generally, cover the arms and necks of infants, especially in Winter. Fathers and mothers, if they were obliged to go with bare arms and necks, even in moderate weather, would often shiver with cold; and yet they have a power of constitution which would subject them to far less hazard and discomfort, than a delicate infant must experience from a similar exposure. This mode of dressing infants, with bare necks and arms, has arisen from the common impression, that they have a power of resisting cold superior to older persons. This is a mistake; for the experiments of medical men have established the fact, that the power of producing heat is least in the period of infancy.

Extensive investigations have been made in France, in reference to this point. It is there required, in some districts, that every infant, at birth, be carried to the office of the *maire,* [*mayor,*] to be registered. It is [Pg 114]found, in these districts, that the deaths of newly-born infants, are much more numerous in the cold, than in the warm, months; and that a much greater proportion of such deaths occurs among those who reside at a distance from the office of the *maire,* than among those in its vicinity. This proves, that exposure to cold has much to do with the continuance of infant life.

But it is as dangerous to go to the other extreme, and keep the body too warm. The skin, when kept at too high a temperature, is relaxed and weakened by too profuse perspiration, and becomes more sensitive, and more readily affected by every change of temperature. This increases the liabilities to sudden colds; and it frequently happens, that the children, who are most carefully guarded from cold, are the ones most liable to take sudden and dangerous chills. The reason is, that, by the too great accumulation of clothing, the skin is too much excited, and the blood is withdrawn from the internal organs, thus weakening them, while the skin itself is debilitated by the same process.

The rule of safety, is, so to cover the body, as to keep it entirely warm, but not so as to induce perspiration in any part. The perspiration induced by exercise is healthful, because it increases the appetite; but the perspiration produced by excess of clothing is debilitating. This shows the importance of adjusting beds and their covering to the season. Featherbeds are unhealthful in warm weather, because they induce perspiration; and in all cases, those, who have the care of children, should proportion their covering by night to the season of the year. Infants and children should never be so clothed, as either to feel chilly, or to induce perspiration.

The greatest trouble, in this respect, to those who have the care of children, is owing to their throwing off their covering in the night. The best guard, against such exposures, is a nightgown, of the warmest and thickest flannel, made like pantaloons at the lower part, and the legs long, so that they can be tied over the feet. [Pg 115]This makes less covering needful, and saves the child from excessive cold when it is thrown off.

The clothing ought always to be proportioned to the constitution and habits. A person of strong constitution, who takes much exercise, needs less clothing than one of delicate and sedentary habits. According to this rule, women need much thicker and warmer clothing, when they go out, than men. But how different are our customs, from what sound wisdom dictates! Women go out with thin stockings, thin shoes, and open necks, when men are protected by thick woollen hose and boots, and their whole body encased in many folds of flannel and broadcloth.

Flannel, worn next the skin, is useful, for several reasons. It is a bad conductor of heat, so that it protects the body from *sudden* chills when in a state of perspiration. It also produces a kind of friction on the skin, which aids it in its functions, while its texture, being loose, enables it to receive and retain much matter, thrown off from the body, which would otherwise accumulate on its surface. This is the reason, why medical men direct, that young children wear flannel next the body, and woollen hose, the first two years of life. They are thus protected from sudden exposures. For the same reason, laboring men should thus wear flannels, which are also considered as preservatives from infection, in unhealthy atmospheres. They give a healthy action to the skin, and thus enable it to resist the operation of unhealthy miasms. On this account, persons residing in a new country should wear such clothing next the skin, to guard them from the noxious miasms caused by extensive vegetable decompositions. It is stated, that the fatal influence of the malaria, or noxious exhalations around Rome, has been much diminished by this practice. But those who thus wear flannel, through the day, ought to take it off, at night, when it is not needed. It should be hung so that it can be well aired, during the night.

But the practice, by which females probably suffer [Pg 116]most, is, the use of *tight dresses*. Much has been said against the use of corsets by ladies. But these may be worn with perfect safety, and be left off, and still injury, such as they often produce, be equally felt. It is the *constriction* of dress, that is to be feared, and not any particular article that produces it. A frock, or a belt, may be so tight, as to be even worse than a corset, which would more equally divide the compression.

So long as it is the fashion to admire, as models of elegance, the wasp-like figures which are presented at the rooms of mantuamakers and milliners, there will be hundreds of foolish women, who will risk their lives and health to secure some resemblance to these deformities of the human frame. But it is believed, that all sensible women, when they fairly understand the evils which result from tight dressing, and learn the *real* model of taste and beauty for a perfect female form, will never risk their own health, or the health of their daughters, in efforts to secure one which is as much at variance with good taste, as it is with good health.

Such female figures as our print-shops present, are made, not by the hand of the Author of all grace and beauty, but by the murderous contrivances of the corset-shop; and the more a woman learns the true rules of grace and beauty for the female form, the more her taste will revolt from such ridiculous distortions. The folly of the Chinese belle, who totters on two useless deformities, is nothing, compared to that of the American belle, who impedes all the internal organs in the discharge of their functions, that she may have a slender waist.

It was shown, in the article on the bones and muscles, that exercise was indispensable to their growth and strength. If any muscles are left unemployed, they diminish in size and strength. The girding of tight dresses operates thus on the muscles of the body. If an article, like corsets, is made to hold up the body, then those muscles, which are designed for this purpose, are released from duty, and grow weak; so that, after [Pg 117]this has been continued for some time, leaving off the unnatural support produces a feeling of weakness. Thus a person will complain of feeling so weak and unsupported, without corsets, as to be uncomfortable. This is entirely owing to the disuse of those muscles, which corsets throw out of employ.

Another effect of tight dress, is, to stop or impede the office of the lungs. Unless the chest can expand, fully, and with perfect ease, a portion of the lungs is not filled with air, and thus the full purification of the blood is prevented. This movement of the lungs, when they are fully inflated, increases the peristaltic movement of the stomach and bowels, and promotes digestion; any constriction of the waist tends to impede this important operation, and indigestion, with all its attendant evils, is often the result.

The rule of safety, in regard to the tightness of dress, is this. Every person should be dressed so loosely, that, *when sitting in the posture used in sewing, reading, or study* THE LUNGS *can be as fully and as easily inflated, as they are without clothing*. Many a woman thinks she dresses loosely, because, when she stands up, her clothing does not confine her chest. This is not a fair test. It is in the position most used when engaged in common employments, that we are to judge of the constriction of dress. Let every woman, then, bear in mind, that, just so long as her dress and position oppose any resistance to the motion of her chest, in just such proportion her blood is unpurified, and her vital organs are debilitated.

The English ladies set our countrywomen a good example, in accommodating their dress to times and seasons. The richest and noblest among them wear warm cotton hose and thick shoes, when they walk for exercise; and would deem it vulgar to appear, as many of our ladies do, with thin hose and shoes, in damp or cold weather. Any mode of dress, not suited to the employment, the age, the season, or the means of the wearer, is in bad taste.

[Pg 118]

CHAPTER IX.
ON CLEANLINESS.

The importance of cleanliness, in person and dress, can never be fully realized, by persons who are ignorant of the construction of the skin, and of the influence which its treatment has on the health of the body. Persons deficient in such knowledge, frequently sneer at what they deem the foolish and fidgety particularity of others, whose frequent ablutions and changes of clothing, exceed their own measure of importance.

The popular maxim, that "dirt is healthy," has probably arisen from the fact, that playing in the open air is very beneficial to the health of children, who thus get dirt on their persons and clothes. But it is the fresh air and exercise, and not the dirt, which promotes the health.

In a previous article, it was shown, that the lungs, bowels, kidneys, and skin, were the organs employed in throwing off those waste and noxious parts of the food not employed in nourishing the body. Of this, the skin has the largest duty to perform; throwing off, at least, twenty ounces every twenty-four hours, by means of insensible perspiration. When exercise sets the blood in quicker motion, it ministers its supplies faster, and there is consequently a greater residuum to be thrown off by the skin; and then the perspiration becomes so abundant as to be perceptible. In this state, if a sudden chill take place, the blood-vessels of the skin contract, the blood is driven from the surface, and the internal organs are taxed with a double duty. If the constitution be a strong one, these organs march on and perform the labor exacted. But if any of these organs be debilitated, the weakest one generally gives way, and some disease ensues.

One of the most frequent illustrations of this reciprocated [Pg 119]action, is afforded by a convivial meeting in cold weather. The heat of the room, the food, and the excitement, quicken the circulation, and perspiration is evolved. When the company passes into the cold air, a sudden revulsion takes place. The increased circulation continues, for some time after; but the skin being cooled, the blood retreats, and the internal organs are obliged to perform the duties of the skin as well as their own. Then, in case the lungs are the weakest organ, the mucous secretion becomes excessive; so that it would fill up the cells, and stop the breathing, were it not for the spasmodic effort called coughing, by which this substance is thrown out. In case the nerves are the weakest part of the system, such an exposure would result in pains in the head or teeth, or in some other nervous ailment. If the muscles be the weakest part, rheumatic affections will ensue; and if the bowels or kidneys be weakest, some disorder in their functions will result.

But it is found, that the closing of the pores of the skin with other substances, tends to a similar result on the internal organs. In this situation, the skin is unable perfectly to perform its functions, and either the blood remains to a certain extent unpurified, or else the internal organs have an unnatural duty to perform. Either of these results tends to produce disease, and the gradual decay of the vital powers.

Moreover, it has been shown, that the skin has the power of absorbing into the blood particles retained on its surface. In consequence of these peculiarities, the skin of the whole body needs to be washed, every day. This process removes from the pores the matter exhaled from the blood, and also that collected from the atmosphere and other bodies. If this process be not often performed, the pores of the skin fill up with the redundant matter expelled, and being pressed, by the clothing, to the surface of the body, the skin is both interrupted in its exhaling process, and its absorbents take back into the system portions of the noxious matter. Thus the [Pg 120]blood is not relieved to the extent designed, while it receives back noxious particles, which are thus carried to the lungs, liver, and every part of the system.

This is the reason why the articles worn next to the skin should often be changed; and why it is recommended that persons should not sleep in the article they wear next the skin through the day. The alternate change and airing of the articles worn next the body by day or night, is a practice very favorable to the health of the skin. The fresh air has the power of removing much of the noxious effluvia received from the body by the clothing. It is with reference to this, that on leaving a bed, its covering should be thrown open and exposed to the fresh air.

The benefit arising from a proper care of the skin, is the reason why bathing has been so extensively practised by civilized nations. The Greeks and Romans considered bathing as indispensable to daily comfort, as much so, as their meals; and public baths were provided for all classes. In European countries, this practice is very prevalent, but there is no civilized nation which pays so little regard to the rules of health, on this subject, as our own. To wash the face, feet, hands, and neck, is the extent of the ablutions practised by perhaps the majority of our people.

In regard to the use of the bath, there is need of some information, in order to prevent danger from its misuse. Persons in good health, and with strong constitutions, can use the cold bath, and the shower-bath, with entire safety and benefit. But if the constitution be feeble, cold bathing is injurious. If it is useful, it can be known by an invigorated feeling, and a warm glow on the skin; but if, instead of this, there be a feeling of debility, and the hands and feet become cold, it is a certain sign, that this kind of bathing is injurious. A bath at ninety-five degrees of Fahrenheit, is about the right temperature. A bath, blood warm, or a little cooler than the skin, is safe for all constitutions, if not protracted over half an hour. After bathing, the body [Pg 121]should be rubbed with a brush or coarse towel, to remove the light scales of scarfskin, which adhere to it, and also to promote a healthful excitement.

A bath should never be taken, till three hours after eating, as it interrupts the process of digestion, by withdrawing the blood from the stomach to the surface. Neither should it be taken, when the body is weary with exercise, nor be immediately followed by severe exercise. Many suppose that a warm bath exposes a person more readily to take cold; and that it tends to debilitate the system. This is not the case, unless it be protracted too long. If it be used so as to cleanse the skin, and give it a gentle stimulus, it is better able to resist cold than before the process. This is the reason why the Swedes and Russians can rush, reeking, out of their steam baths, and throw themselves into the snow, and not only escape injury, but feel invigorated. It is for a similar reason, that we suffer less in going into the cold, from a warm room, with our body entirely warm, than when we go out somewhat chilled. When the skin is warm, the circulation is active on the surface, and the cold does not so reduce its temperature, but that increased exercise will keep up its warmth.

When families have no bathing establishment, every member should wash the whole person, on rising or going to bed, either in cold or warm water, according to the constitution. It is especially important, that children have the perspiration and other impurities, which their exercise and sports have occasioned, removed from their skin before going to bed. The hours of sleep are those when the body most freely exhales the waste matter of the system, and all the pores should be properly freed from impediments to this healthful operation. For this purpose, a large tin wash-pan should be kept for children, just large enough, at bottom, for them to stand in, and flaring outward, so as to be very broad at top. A child can then be placed in it, standing, and washed with a sponge, without wetting the floor. Being small at bottom, it is better than a tub; it is not only smaller, but lighter, and requires less water.

[Pg 122]

These remarks indicate the wisdom of those parents, who habitually wash their children, all over, before they go to bed. The chance of life and health, to such children, is greatly increased by this practice; and no doubt much of the suffering of childhood, from cutaneous eruptions, weak eyes, earache, colds, and fevers, is owing to a neglect of the skin.

The care of the teeth should be made habitual to children, not merely as promoting an agreeable appearance, but as a needful preservative. The saliva contains tartar, an earthy substance, which is deposited on the teeth, and destroys both their beauty and health. This can be prevented, by the use of the brush, night and morning. But, if this be neglected, the deposite becomes hard, and can be removed only by the dentist. If suffered to remain, it tends to destroy the health of the gums; they gradually decay, and thus the roots of the teeth become bare, and they often drop out.

When children are shedding their first set of teeth, care should be taken, to remove them as soon as they become loose; otherwise the new teeth will grow awry. When persons have defective teeth, they can often be saved, by having them filled by a dentist. This also will frequently prevent the toothache.

Children should be taught to take proper care of their nails. Long and dirty nails have a disagreeable appearance. When children wash, in the morning, they should be supplied with an instrument to clean the nails, and be required to use it.

CHAPTER X.
ON EARLY RISING.

There is no practice, which has been more extensively eulogized, in all ages, than early rising; and this universal impression, is an indication that it is founded [Pg 123]on true philosophy. For, it is rarely the case, that the common sense of mankind fastens on a practice, as really beneficial, especially one that demands self-denial, without some substantial reason.

This practice, which may justly be called a domestic virtue, is one, which has a peculiar claim to be styled American and democratic. The distinctive mark of aristocratic nations, is, a disregard of the great mass, and a disproportionate regard for the interests of certain privileged orders. All the customs and habits of such a nation, are, to a greater or less extent, regulated by this principle. Now the mass of any nation must always consist of persons who labor at occupations which demand the light of day. But in aristocratic countries, especially in England, labor is regarded as the mark of the lower classes, and indolence is considered as one mark of a gentleman. This impression has gradually and imperceptibly, to a great extent, regulated their customs, so that, even in their hours of meals and repose, the higher orders aim at being different and distinct from those, who, by laborious pursuits, are placed below them. From this circumstance, while the lower orders labor by day, and sleep at night, the rich, the noble, and the honored, sleep by day, and follow their pursuits and pleasures by night. It will be found, that the aristocracy of London breakfast near mid-day, dine after dark, visit and go to Parliament between ten and twelve at night, and retire to sleep towards morning. In consequence of this, the subordinate classes, who aim at gentility, gradually fall into the same practice. The influence of this custom extends across the ocean, and here, in this democratic land, we find many, who measure their grade of gentility by the late hour at which they arrive at a party. And this aristocratic tendency is growing upon us, so that, throughout the Nation, the hours for visiting and retiring are constantly becoming later, while the hours for rising correspond in lateness.

The question, then, is one which appeals to American [Pg 124]women, as a matter of patriotism; as having a bearing on those great principles of democracy, which we conceive to be equally the principles of Christianity. Shall we form our customs on the principle that labor is degrading, and indolence genteel? Shall we assume, by our practice, that the interests of the great mass are to be sacrificed for the pleasures and honors of a privileged few? Shall we ape the customs of aristocratic lands, in those very practices which result from principles and institutions that we condemn? Shall we not rather take the place to which we are entitled, as the leaders, rather than the followers, in the customs of society, turn back the tide of aristocratic inroads, and carry through the whole, not only of civil and political, but of social and domestic, life, the true principles of democratic freedom and equality? The following considerations may serve to strengthen an affirmative decision.

The first, relates to the health of a family. It is a universal law of physiology, that all living things flourish best in the light. Vegetables, in a dark cellar, grow pale and spindling,[H] and children, brought up in mines, are wan and stinted. This universal law, indicates the folly of turning day into night, thus losing the genial influence, which the light of day produces on all animated creation.

There is another phenomenon in the physiology of Nature, which equally condemns this practice. It has been shown, that the purification of the blood, in the lungs, is secured, by the oxygen of the atmosphere absorbing its carbon and hydrogen. This combination forms carbonic acid and water, which are expired from our lungs into the atmosphere. Now all the vegetable world undergoes a similar process. In the light of day, all the leaves of vegetables absorb carbon and expire oxygen, thus supplying the air with its vital principle, and withdrawing the more deleterious element. But, when the light is withdrawn, this process is reversed, [Pg 125]and all vegetables exhale carbonic acid, and inspire the oxygen of the air. Thus it appears, that the atmosphere of day is much more healthful than that of the night, especially out of doors.

Moreover, when the body is fatigued, it is much more liable to deleterious influences, from noxious particles in the atmosphere, which may be absorbed by the skin or the lungs. In consequence of this, the last hours of daily labor are more likely to be those of risk, especially to delicate constitutions. This is a proper reason for retiring to the house and to slumber, at an early hour, that the body may not be exposed to the most risk, when, after the exertions of the day, it is least able to bear it.

The observations of medical men, whose inquiries have been directed to this point, have decided, that from six to eight hours, is the amount of sleep demanded by persons in health. Some constitutions require as much as eight, and others no more than six, hours of repose. But eight hours is the maximum for all persons in ordinary health, with ordinary occupations. In cases of extra physical exertions, or the debility of disease, or a decayed constitution, more than this is required. Let eight hours, then, be regarded as the ordinary period required for sleep, by an industrious people, like the Americans. According to this, the practice of rising between four and five, and retiring between nine and ten, in Summer, would secure most of the sunlight, and expose us the least to that period of the atmosphere, when it is most noxious. In Winter, the night air is less deleterious, because the frost binds noxious exhalations, and vegetation ceases its inspiring and expiring process; and, moreover, as the constitution is more tried, in cold, than in warm, weather, and as in cold weather the body exhales less during the hours of sleep, it is not so injurious to protract our slumbers beyond the proper period, as it is in the warm months. But in Winter, it is best for grown persons, in health, to rise as soon as they can see to [Pg 126]dress, and retire so as not to allow more than eight hours for sleep.

It thus appears, that the laws of our political condition, the laws of the natural world, and the constitution of our bodies, alike demand that we rise with the light of day to prosecute our employments, and that we retire within doors, when this light is withdrawn.

In regard to the effects of protracting the time spent in repose, many extensive and satisfactory investigations have been made. It has been shown, that, during sleep, the body perspires most freely, while yet neither food nor exercise are ministering to its wants. Of course, if we continue our slumbers, beyond the time required to restore the body to its usual vigor, there is an unperceived undermining of the constitution, by this protracted and debilitating exhalation. This process, in a course of years, renders the body delicate, and less able to withstand disease; and in the result shortens life. Sir John Sinclair, who has written a large work on the Causes of Longevity, states, as one result of his extensive investigations, that he has never yet heard or read of a single case of great longevity, where the individual was not an early riser. He says, that he has found cases, in which the individual has violated some one of all the other laws of health, and yet lived to great age; but never a single instance, in which any constitution has withstood that undermining, consequent on protracting the hours of repose beyond the demands of the system.

Another reason for early rising, is, that it is indispensable to a systematic and well-regulated family. At whatever hour the parents retire, children and domestics, wearied by play or labor, must retire early. Children usually awake with the dawn of light, and commence their play, while domestics usually prefer the freshness of morning for their labors. If, then, the parents rise at a late hour, they either induce a habit of protracting sleep in their children and domestics, or else the family is up, and at their pursuits, while their [Pg 127]supervisors are in bed. Any woman, who asserts that her children and domestics, in the first hours of day, when their spirits are freshest, will be as well regulated without her presence, as with it, confesses that, which surely is little for her credit. It is believed, that any candid woman, whatever may be her excuse for late rising, will concede, that, if she could rise early, it would be for the advantage of her family. A late breakfast puts back the work, through the whole day, for every member of a family; and, if the parents thus occasion the loss of an hour or two, to each individual, who, but for their delay in the morning, would be usefully employed, they, alone, are responsible for all this waste of time. Is it said, that those, who wish to rise early, can go to their employments before breakfast? it may be replied, that, in most cases, it is not safe to use the eyes or the muscles in the morning, till the losses of the night have been repaired by food. In addition to this, it may be urged, that, where the parents set an example of the violation of the rules of health and industry, their influence tends in the wrong direction;

136

so that whatever waste of time is induced, by a practice which they thus uphold, must be set down to their account.

But the practice of early rising has a relation to the general interests of the social community, as well as to that of each distinct family. All that great portion of the community, who are employed in business and labor, find it needful to rise early; and all their hours of meals, and their appointments for business or pleasure, must be accommodated to these arrangements. Now, if a small portion of the community establish very different hours, it makes a kind of jostling, in all the concerns and interests of society. The various appointments for the public, such as meetings, schools, and business hours, must be accommodated to the mass, and not to individuals. The few, then, who establish domestic habits at variance with the majority, are either constantly interrupted in their own arrangements, or [Pg 128]else are interfering with the rights and interests of others. This is exemplified in the case of schools. In families where late rising is practised, either hurry, irregularity, and neglect, are engendered in the family, or else the interests of the school, and thus of the community, are sacrificed. In this, and many other concerns, it can be shown, that the wellbeing of the bulk of the people, is, to a greater or less extent, impaired by this aristocratic practice. Let any teacher select the unpunctual scholars,—a class who most seriously interfere with the interests of the school;—and let men of business select those who cause them most waste of time and vexation, by unpunctuality; and it will be found, that they are among the late risers, and rarely among those who rise early. Thus, it is manifest, that late rising not only injures the person and family which practise it, but interferes with the rights and convenience of the community.

FOOTNOTE:

[H] Shooting into a long, small, stalk or root.

CHAPTER XI.
ON DOMESTIC EXERCISE.

In the preceding chapters, we have noticed the various causes, which, one or all, operate to produce that melancholy delicacy and decay of the female constitution, which are the occasion of so much physical and mental suffering throughout this Country.

These, in a more condensed form, may be enumerated thus:

A want of exercise, inducing softness in the bones, weakness in the muscles, inactivity in the digestive organs, and general debility in the nervous system: A neglect of the care of the skin, whereby the blood has not been properly purified, and the internal organs have been weakened: A violation of the laws of health, in regard to food, by eating too much, too fast, and too [Pg 129]often; by using stimulating food and drinks; by using them too warm or too cold; and by eating that which the power of the stomach is not sufficient to digest: A neglect of the laws of health, in regard to clothing, by dressing too tight, and by wearing too little covering, in cold and damp weather, and especially by not sufficiently protecting the feet: A neglect to gain a proper supply of pure air, in sleeping apartments and schoolrooms, and too great a confinement to the house: The pursuit of exciting amusements at unseasonable hours, and the many exposures involved at such times: And lastly, sleeping by day, instead of by night, and protracting the hours of sleep, beyond the period of repose demanded for rest; thus exhausting, instead of recruiting, the energies of the system.

But all the other causes, combined, probably, do not produce one half the evils, which result from a want of proper exercise. A person who keeps all the functions of the system in full play, by the active and frequent use of every muscle, especially if it be in the open air, gains a power of constitution, which can resist many evils that would follow from the other neglects and risks detailed. This being the case, there can be no subject, more important for mothers and young ladies to understand, than the influence on the health, both of body and mind, of the neglect or abuse of the muscular system.

It has been shown, in the previous pages, that all the muscles have nerves and blood-vessels, running in larger trunks, or minute branches, to every portion of the body. The experiments of Sir Charles Bell and others, have developed the curious fact, that each apparently single nerve, in reality consists of two distinct portions, running together in the same covering. One portion, is the nerve of *sensation* or *feeling*, the other, the nerve of *motion*. The nerves of sensation are those which are affected by the emotions and volitions of the mind; and the nerves of motion are those which impart moving power to the muscles. Experiments show, [Pg 130]that, where the nerves issue from the spine, the nerve of sensation may be cut off without severing the nerve of motion, and then the parts, to which this nerve extends, lose the power of feeling, while the power of motion continues; and so, on the other hand, the nerve of motion may be divided, and, the nerve of sensation remaining uninjured, the power of feeling is retained, and the power of motion is lost.

In certain nervous diseases, sometimes a limb loses its power of feeling, and yet retains the power of motion; in other cases, the power of motion is lost, and the power of sensation is retained; and in other cases, still, when a limb is *paralysed*, both the power of motion and of sensation are lost.

Now, the nerves, like all other parts of the body, gain and lose strength, according as they are exercised. If they have too much, or too little, exercise, they lose strength; if they are exercised to a proper degree, they gain strength. When the mind is continuously excited, by business, study, or the imagination, the nerves of feeling are kept in constant action, while the nerves of motion are unemployed. If this is continued, for a long time, the nerves of sensation lose their strength, from over action, and the nerves of motion lose their power, from inactivity. In consequence, there is a morbid excitability of the nervous, and a debility of the muscular, system, which make all exertion irksome and wearisome. The only mode of preserving the health of these systems, is, to keep up in them an equilibrium of action. For this purpose, occupations must be sought, which exercise the muscles, and interest the mind; and thus the equal action of both kinds of nerves is secured. This shows why exercise is so much more healthful and invigorating, when the mind is interested, than when it is not. As an illustration, let a person go a shopping, with a friend, and have nothing to do, but look on; how soon do the continuous walking and standing weary! But suppose one, thus wearied, hears of the arrival of a very dear friend: she can instantly walk off a mile or[Pg 131] two, to meet her, without the least feeling of fatigue. By this is shown the importance of furnishing, for young persons, exercise in which they will take an interest. Long and formal walks, merely for exercise, though they do some good, in securing fresh air and some exercise of the muscles, would be of triple benefit, if changed to amusing sports, or to the cultivation of fruits and flowers, in which it is impossible to engage, without acquiring a great interest. It shows, also, why it is far better to trust to useful domestic exercise, at home, than to send a young person out to walk, for the mere purpose of exercise. Young girls can seldom be made to realize the value of health, and the need of exercise to secure it, so as to feel much interest in walking abroad, when they have no other object. But, if they are brought up to minister to the comfort and enjoyment of themselves and others, by performing domestic duties, they will constantly be interested and cheered in their exercise, by the feeling of usefulness, and the consciousness of having performed their duty.

There are few young persons, it is hoped, who are brought up with such miserable habits of selfishness and indolence, that they cannot be made to feel happier, by the consciousness of being usefully employed. And those who have never been accustomed to think or care for any one but themselves, and who seem to feel little pleasure in making themselves useful, by wise and proper influences, can often be gradually awakened to the new pleasure of benevolent exertion to promote the comfort and enjoyment of others. And the more this sacred and elevating kind of enjoyment is tasted, the greater is the relish induced. Other enjoyments, often cloy; but the heavenly pleasure, secured by virtuous industry and benevolence, while it satisfies, at the time, awakens fresh desires for so ennobling a good.

But, besides the favorable influence on the nervous and muscular system, thus gained, it has been shown, that exercise imparts fresh strength and vitality to all parts of the body. The exertion of the muscles quickens [Pg 132]the flow of the blood, which thus ministers its supplies faster to every part of the body, and, of course, loses a portion of its nourishing qualities. When this is the case, the stomach issues its mandate of *hunger*, calling for new supplies. When these are furnished, the action of the muscles again hastens a full supply to every organ, and thus the nerves, the muscles, the bones, the skin, and all the internal organs, are invigorated, and the whole body developes its powers, in fair proportions, fresh strength and full beauty. All the cosmetics of trade, all the labors of mantuamakers, milliners, makers of corsets, shoemakers, and hairdressers, could never confer so clear and pure a skin, so fresh a color, so finely moulded a form, and such cheerful health and spirits, as would be secured by training a child to obey the laws of the benevolent Creator, in the appropriate employment of body and mind in useful domestic exercise. And the present habits of the wealthy, and even of those without wealth, which condemn young girls so exclusively to books or sedentary pursuits, are as destructive to beauty and grace, as they are to health and happiness.

Every allowance should be made for the mistakes of mothers and teachers, to whom the knowledge which would have saved them from the evils of such a course has never been furnished; but as information, on these matters, is every year becoming more abundant, it is to be hoped, that the next generation, at least, may be saved from the evils which afflict those now on the stage. What a change would be made in the happiness of this Country, if all the pale and delicate young girls should become blooming, healthful, and active, and all the enfeebled and care-worn mothers should be transformed into such fresh, active, healthful, and energetic matrons, as are so frequently found in our mother land!

It has been stated, that the excessive use of the muscles, as much as their inactivity, tends to weaken them. Nothing is more painful, than the keeping a muscle constantly on the stretch, without any relaxation [Pg 133]or change. This can be realized, by holding out an arm, perpendicularly to the body, for ten or fifteen minutes, if any one can so long bear the pain. Of course, confinement to one position, for a great length of time, tends to weaken the muscles thus strained.

This shows the evil of confining young children to their seats, in the schoolroom, so much and so long as is often done. Having no backs to their seats, as is generally the case, the muscles, which are employed in holding up the body, are kept in a state of constant tension, till they grow feeble from overworking. Then, the child begins to grow crooked, and the parents, to remedy the evil, sometimes put on bracers or corsets. These, instead of doing any good, serve to prevent the use of those muscles, which, if properly exercised, would hold the body straight; and thus they grow still weaker, from entire inactivity. If a parent perceives that a child is growing crooked, the proper remedy is, to withdraw it from all pursuits which tax one particular set of muscles, and turn it out to exercise in sports, or in gardening, in the fresh air, when all the muscles will be used, and the whole system strengthened. Or, if this cannot be done, sweeping, dusting, running of errands, and many household employments, which involve lifting, stooping, bending, and walking, are quite as good, and, on some accounts, better, provided the house is properly supplied with fresh air.

Where persons have formed habits of inactivity, some caution is necessary, in attempting a change; this must be made gradually; and the muscles must never be excessively fatigued at any time. If this change be not thus gradually made, the weakness, at first caused by inactivity, will be increased by excessive exertion. A distinguished medical gentleman gives this rule, to direct us in regard to the amount of fatigue, which is safe and useful. A person is never too much fatigued, if one night of repose gives sufficient rest, and restores the usual strength. But, if the sleep is disturbed, and the person wakes with a feeling[Pg 134] of weariness and languor, it is a sure indication that the exercise has been excessive. No more fatigue, then, should be allowed, than one night's rest will remedy.

Some persons object to sweeping, on account of the dust inhaled. But free ventilation, frequent sweeping, and the use of damp sand, or damp Indian meal, or damp tea leaves, for carpets, will secure a more clear atmosphere than is often found in the streets of cities. And the mother, who will hire domestics, to take away this and other domestic employments, which would secure to her daughters, health, grace, beauty, and domestic virtues, and the young ladies, who consent to be deprived of these advantages, will probably live to mourn over the languor, discouragement, pain, and sorrow, which will come with ill health, as the almost inevitable result.

The following are extracts from 'The Young Ladies' Friend,' on this subject: —

"Whether rich or poor, young or old, married or single, a woman is always liable to be called to the performance of every kind of domestic duty, as well as to be placed at the head of a family; and nothing, short of a *practical* knowledge of the details of housekeeping, can ever make those duties easy, or render her competent to direct others in the performance of them.

"All moral writers on female character, treat of Domestic Economy as an indispensable part of female education; and this, too, in the old countries of Europe, where an abundant population, and the institutions of society, render it easy to secure the services of faithful domestics."

"All female characters that are held up to admiration, whether in fiction or biography, will be found to possess these domestic accomplishments; and, if they are considered indispensable in the Old World, how much more are they needed, in this land of independence, where riches cannot exempt the mistress of[Pg 135] a family from the difficulty of procuring efficient aid, and where perpetual change of domestics, renders perpetual instruction and superintendence necessary.

"Since, then, the details of good housekeeping must be included in a good female education, it is very desirable that they should be acquired when young, and so practised as to become easy, and to be performed dexterously and expeditiously."

"The elegant and accomplished Lady Mary Wortley Montagu, who figured in the fashionable, as well as the literary, circles of her time, has said, that 'the most minute details of household economy become elegant and refined, when they are ennobled by sentiment;' and they are truly ennobled, when we do them either from a sense of duty, or consideration for a parent, or love to a husband. 'To furnish a room,' continues this lady, 'is no longer a commonplace affair, shared with upholsterers and cabinet-makers; it is decorating the place where I am to meet a friend or lover. To order dinner is not merely arranging a meal with my cook; it is preparing refreshment for him whom I love. These necessary occupations, viewed in this light, by a person capable of strong attachment, are so many pleasures, and afford her far more delight, than the games and shows which constitute the amusements of the world.'

"Such is the testimony of a titled lady of the last century, to the sentiment that may be made to mingle in the most homely occupations. I will now quote that of a modern female writer and traveller, who, in her pleasant book, called 'Six Weeks on the Loire,' has thus described the housewifery of the daughter of a French nobleman, residing in a superb chateau on that river. The travellers had just arrived, and been introduced, when the following scene took place.

"'The bill of fare for dinner was discussed in my presence, and settled, *sans façon*,[I] with that delightful [Pg 136]frankness and gayety, which, in the French character, gives a charm to the most trifling occurrence. Mademoiselle Louise then begged me to excuse her for half an hour, as she was going to make some creams, and some *pastilles*.[J] I requested that I might accompany her, and also render myself useful; we accordingly went together to the dairy. I made tarts *à l'Anglaise*,[K] whilst she made confections and *bonbons*,[L] and all manner of pretty things, with as much ease as if she had never done any thing else, and as much grace as she displayed in the saloon. I could not help thinking, as I looked at her, with her servants about her, all cheerful, respectful, and anxious to attend upon her, how much better it would be for the young ladies in England, if they would occasionally return to the habits of their grandmammas, and mingle the animated and endearing occupations of domestic life, and the modest manners and social amusements of home, with the perpetual practising on harps and pianos, and the incessant efforts at display, and search after gayety, which, at the present day, render them any thing but what an amiable man, of a reflecting mind and delicate sentiments, would desire in the woman he might wish to select as the companion of his life.'"

FOOTNOTES:

[I] Without formality, or useless ceremony.

[J] Rolls of paste, or pastry, or sugarplums.

[K] According to the English fashion.

[L] Nice things or dainties, such as sweetmeats.

CHAPTER XII.
ON DOMESTIC MANNERS.

Good-manners are the expressions of benevolence in personal intercourse, by which we endeavor to promote the comfort and enjoyment of others, and to avoid all that gives needless uneasiness. It is the exterior exhibition of the Divine precept, which requires us to do[Pg 137] to others, as we would that they should do to us. It is saying, by our deportment, to all around, that we consider their feelings, tastes, and convenience, as equal in value to our own.

Good-manners lead us to avoid all practices which offend the taste of others; all violations of the conventional rules of propriety; all rude and disrespectful language and deportment; and all remarks, which would tend to wound the feelings of another.

There is a serious defect, in the manners of the American people, especially in the free States, which can never be efficiently remedied, except in the domestic circle, and during early life. It is a deficiency in the free expression of kindly feelings and sympathetic emotions, and a want of courtesy in deportment. The causes, which have led to this result, may easily be traced.

The forefathers of this Nation, to a wide extent, were men who were driven from their native land, by laws and customs which they believed to be opposed both to civil and religious freedom. The sufferings they were called to endure, the subduing of those gentler feelings which bind us to country, kindred, and home, and the constant subordination of the passions to stern principle, induced characters of great firmness and self-control. They gave up the comforts and refinements of a civilized country, and came, as pilgrims, to a hard soil, a cold clime, and a heathen shore. They were continually forced to encounter danger, privations, sickness, loneliness, and death; and all these, their religion taught them to meet with calmness, fortitude, and submission. And thus it became the custom and habit of the whole mass, to repress, rather than to encourage, the expression of feeling.

Persons who are called to constant and protracted suffering and privation, are forced to subdue and conceal emotion; for the free expression of it would double their own suffering, and increase the sufferings of others. Those, only, who are free from care and[Pg 138] anxiety, and whose minds are mainly occupied by cheerful emotions, are at full liberty to unveil their feelings.

It was under such stern and rigorous discipline, that the first children in New England were reared; and the manners and habits of parents are usually, to a great extent, transmitted to children. Thus it comes to pass, that the descendants of the Puritans, now scattered over every part of the Nation, are predisposed to conceal the gentler emotions, while their manners are calm, decided, and cold, rather than free and impulsive. Of course, there are very many exceptions to these predominating results.

The causes, to which we may attribute a general want of courtesy in manners, are certain incidental results of our democratic institutions. Our ancestors, and their descendants, have constantly been combating the aristocratic principle, which would exalt one class of men at the expense of another. They have had to contend with this principle, not only in civil, but in social, life. Almost every American, in his own person, as well as in behalf of his class, has had to assume and defend the main principle of democracy,—that every man's feelings and interests are equal in value to those of every other man. But, in doing this, there has been some want of clear discrimination. Because claims, based on distinctions of mere birth, fortune, or position, were found to be injurious, many have gone to the extreme of inferring that all distinctions, involving subordination, are useless. Such, would regard children as equals to parents, pupils to teachers, domestics to their employers, and subjects to magistrates; and that, too, in all respects.

The fact, that certain grades of superiority and subordination are needful, both for individual and public benefit, has not been clearly discerned; and there has been a gradual tendency to an extreme, which has sensibly affected our manners. All the proprieties and courtesies, which depend on the recognition of the[Pg 139] relative duties of superior and subordinate, have been warred upon; and thus we see, to an increasing extent, disrespectful treatment of parents, from children; of teachers, from pupils; of employers, from domestics; and of the aged, from the young. In all classes and circles, there is a gradual decay in courtesy of address.

In cases, too, where kindness is rendered, it is often accompanied with a cold, unsympathizing manner, which greatly lessens its value, while kindness or politeness is received in a similar style of coolness, as if it were but the payment of a just due.

It is owing to these causes, that the American people, especially the inhabitants of New England, do not do themselves justice. For, while those, who are near enough to learn their real character and feelings, can discern the most generous impulses, and the most kindly sympathies, they are so veiled, in a composed and indifferent demeanor, as to be almost entirely concealed from strangers.

These defects in our national manners, it especially falls to the care of mothers, and all who have charge of the young, to rectify; and if they seriously undertake the matter, and wisely adapt means to ends, these defects will be remedied. With reference to this object, the following ideas are suggested.

The law of Christianity and of democracy, which teaches that all men are born equal, and that their interests and feelings should be regarded as of equal value, seems to be adopted in aristocratic circles, with exclusive reference to the class in which the individual moves. The courtly gentleman, addresses all of his own class with politeness and respect; and, in all his actions, seems to allow that the feelings and convenience of others are to be regarded, the same as his own. But his demeanor to those of inferior station, is not based on the same rule.

Among those, who make up aristocratic circles, such as are above them, are deemed of superior, and such as are below, of inferior, value. Thus, if a young,[Pg 140] ignorant, and vicious coxcomb, happens to be born a lord, the aged, the virtuous, the learned, and the wellbred, of another class, must give his convenience the precedence, and must address him in terms of respect. So, when a man of noble birth is thrown among the lower classes, he demeans himself in a style, which, to persons of his own class, would be deemed the height of assumption and rudeness.

Now, the principles of democracy require, that the same courtesy, which we accord to our own circle, shall be extended to every class and condition; and that distinctions, of superiority and subordination, shall depend, not on accidents of birth, fortune, or occupation, but solely on those relations, which the good of all classes equally require. The distinctions demanded, in a democratic state, are simply those, which result from relations, that are common to every class, and are for the benefit of all.

It is for the benefit of every class, that children be subordinate to parents, pupils to teachers, the employed to their employers, and subjects to magistrates. In addition to this, it is for the general well-being, that the comfort or convenience of the delicate and feeble, should be preferred to that of the strong and healthy, who would suffer less by any deprivation, and that precedence should be given to their elders, by the young, and that reverence should be given to the hoary head.

The rules of good-breeding, in a democratic state, must be founded on these principles. It is, indeed, assumed, that the value of the happiness of each individual, is the same as that of every other; but, as there must be occasions, where there are advantages which all cannot enjoy, there must be general rules for regulating a selection. Otherwise, there would be constant scrambling, among those of equal claims, and brute force must be the final resort; in which case the strongest would have the best of every thing. The democratic rule, then, is, that superiors,[Pg 141] in age, station, or office, have precedence of subordinates; age and feebleness, of youth and strength; and the feebler sex, of more vigorous man.[M]

There is, also, a style of deportment and address, which is appropriate to these different relations. It is suitable for a superior to secure compliance with his wishes, from those subordinate to him, by commands; but a subordinate must secure compliance with his wishes, from a superior, by requests. It is suitable for a parent, teacher, or employer, to admonish for neglect of duty; but not for an inferior to adopt such a course towards a superior. It is suitable for a superior to take precedence of a subordinate, without any remark; but not for an inferior, without previously asking leave, or offering an apology. It is proper for a superior to use language and manners of freedom and familiarity, which would be improper from a subordinate to a superior.

The want of due regard to these proprieties, occasions the chief defect in American manners. It is very common to hear children talk to their parents, in a style proper only between companions and equals; so, also, the young address their elders, those employed, their employers, and domestics, the members of the family and their visiters, in a style, which is inappropriate to their relative positions. A respectful address is required not merely towards superiors; every person desires to be treated with courtesy and respect, and therefore, the law of benevolence demands such demeanor, towards all whom we meet in the social intercourse of life. "Be ye courteous," is the direction of the Apostle in reference to our treatment of *all*.

Good-manners can be successfully cultivated, only in early life, and in the domestic circle. There is nothing which depends so much upon *habit*, as the constantly [Pg 142]recurring proprieties of good-breeding; and, if a child grows up without forming such habits, it is very rarely the case that they can be formed at a later period. The feeling, that it is of little consequence how we behave at home, if we conduct properly abroad, is a very fallacious one. Persons, who are careless and ill bred at home, may imagine that they can assume good-manners abroad; but they mistake. Fixed habits of tone, manner, language, and movements, cannot be suddenly altered; and those who are illbred at home, even when they try to hide their bad habits, are sure to violate many of the obvious rules of propriety, and yet be unconscious of it.

And there is nothing, which would so effectually remove prejudice against our democratic institutions, as the general cultivation of good-breeding in the domestic circle. Good-manners are the exterior of benevolence, the minute and often recurring exhibitions of "peace and good-will;" and the nation, as well as the individual, which most excels in the external, as well as the internal, principle, will be most respected and beloved.

The following are the leading points, which claim attention from those who have the care of the young.

In the first place, in the family, there should be required, a strict attention to the rules of precedence, and those modes of address appropriate to the various relations to be sustained. Children should always be required to offer their superiors, in age or station, the precedence in all comforts and conveniences, and always address them in a respectful tone and manner. The custom of adding "Sir," or "Ma'am," to "Yes," or "No," is valuable, as a perpetual indication of a respectful recognition of superiority. It is now going out of fashion, even among the most wellbred people; probably from a want of consideration of its importance. Every remnant of courtesy of address, in our customs, should be carefully cherished, by all who feel a value for the proprieties of good-breeding.

[Pg 143]

If parents allow their children to talk to them, and to the grown persons in the family, in the same style in which they address each other, it will be vain to hope for the courtesy of manner and tone, which good-breeding demands in the general intercourse of society. In a large family, where the elder children are grown up, and the younger are small, it is important to require the latter to treat the elder as superiors. There are none, so ready as young children to assume airs of equality; and, if they are allowed to treat one class of superiors in age and character disrespectfully, they will soon use the privilege universally. This is the reason, why the youngest children of a family are most apt to be pert, forward, and unmannerly.

Another point to be aimed at, is, to require children always to acknowledge every act of kindness and attention, either by words or manner. If they are so trained as always to make grateful acknowledgements, when receiving favors, one of the objectionable features in American manners will be avoided.

Again, children should be required to ask leave, whenever they wish to gratify curiosity, or use an article which belongs to another. And if cases occur, when they cannot comply with the rules of good-breeding, as, for instance, when they must step between a person and the fire, or take the chair of an older person, they should be required either to ask leave, or to offer an apology.

There is another point of good-breeding, which cannot, in all cases, be understood and applied by children, in its widest extent. It is that, which requires us to avoid all remarks which tend to embarrass, vex, mortify, or in any way wound the feelings, of another. To notice personal defects; to allude to others' faults, or the faults of their friends; to speak disparagingly of the sect or party to which a person belongs; to be inattentive, when addressed in conversation; to contradict flatly; to speak in contemptuous tones of opinions expressed by another;—all these, are violations of the[Pg 144] rules of good-breeding, which children should be taught to regard. Under this head, comes the practice of whispering, and staring about, when a teacher, or lecturer, or clergyman, is addressing a class or audience. Such inattention, is practically saying, that what the person is uttering is not worth attending to; and persons of real good-breeding always avoid it. Loud talking and laughing, in a large assembly, even when no exercises are going on; yawning and gaping in company; and not looking in the face a person who is addressing you, are deemed marks of ill-breeding.

Another branch of good-manners, relates to the duties of hospitality. Politeness requires us to welcome visiters with cordiality; to offer them the best accommodations; to address conversation to them; and to express, by tone and manner, kindness and respect. Offering the hand to all visiters, at one's own house, is a courteous and hospitable custom; and a cordial shake of the hand, when friends meet, would abate much of the coldness of manner ascribed to Americans.

The last point of good-breeding, to be noticed, refers to the conventional rules of propriety and good taste. Of these, the first class relates to the avoidance of all disgusting or offensive personal habits, such as fingering the hair; cleaning the teeth or nails; picking the nose; spitting on carpets; snuffing, instead of using a handkerchief, or using the article in an offensive manner; lifting up the boots or shoes, as some men do, to tend them on the knee, or to finger them;—all these tricks, either at home or in society, children should be taught to avoid.

Another branch, under this head, may be called *table manners*. To persons of good-breeding, nothing is more annoying, than violating the conventional proprieties of the table. Reaching over another person's plate; standing up, to reach distant articles, instead of asking to have them passed; using one's own knife, and spoon, for butter, salt, or sugar, when it is the custom of the family to provide separate utensils for the[Pg 145] purpose; setting cups, with tea dripping from them, on the tablecloth, instead of the mats or small plates furnished; using the tablecloth, instead of the napkins; eating fast, and in a noisy manner; putting large pieces in the mouth; looking and eating as if very hungry, or as if anxious to get at certain dishes; sitting at too great a distance from the table, and dropping food; laying the knife and fork on the tablecloth, instead of on the bread, or the edge of the plate;—all these particulars, children should be taught to avoid. It is always desirable, too, to require children, when at table with grown persons, to be silent, except when addressed by others; or else their chattering will interrupt the conversation and comfort of their elders. They should always be required, too, to wait, *in silence,* till all the older persons are helped.

All these things should be taught to children, gradually, and with great patience and gentleness. Some parents, with whom good-manners is a great object, are in danger of making their children perpetually uncomfortable, by suddenly surrounding them with so many rules, that they must inevitably violate some one or other, a great part of the time. It is much better to begin with a few rules, and be steady and persevering with these, till a habit is formed, and then take a few more, thus making the process easy and gradual. Otherwise, the temper of children will be injured; or, hopeless of fulfilling so many requisitions, they will become reckless and indifferent to all.

But, in reference to those who have enjoyed advantages for the cultivation of good-manners, and who duly estimate its importance, one caution is necessary. Those, who never have had such habits formed in youth, are under disadvantages, which no benevolence of temper can remedy. They may often violate the tastes and feelings of others, not from a want of proper regard for them, but from ignorance of custom, or want of habit, or abstraction of mind, or from other causes, which demand forbearance and sympathy, rather than[Pg 146] displeasure. An ability to bear patiently with defects in manners, and to make candid and considerate allowance for a want of advantages, or for peculiarities in mental habits, is one mark of the benevolence of real good-breeding.

The advocates of monarchical and aristocratic institutions, have always had great plausibility given to their views, by the seeming tendencies to insubordination and bad-manners, of our institutions. And it has been too indiscriminately conceded, by the defenders of the latter, that such are these tendencies, and that the offensive points, in American manners, are the necessary result of democratic principles.

But it is believed, that both facts and reasoning are in opposition to this opinion. The following extract from the work of De Tocqueville, exhibits the opinion of an impartial observer, when comparing American manners with those of the English, who are confessedly the most aristocratic of all people.

He previously remarks on the tendency of aristocracy to make men more sympathizing with persons of their own peculiar class, and less so towards those of lower degree; and he then contrasts American manners with the English, claiming that the Americans are much the most affable, mild, and social. "In America, where the privileges of birth never existed, and where riches confer no peculiar rights on their possessors, men acquainted with each other are very ready to frequent the same places, and find neither peril nor advantage in the free interchange of their thoughts. If they meet, by accident, they neither seek nor avoid intercourse; their manner is therefore natural, frank, and open." "If their demeanor is often cold and serious, it is never haughty or constrained." But an "aristocratic pride is still extremely great among the English; and, as the limits of aristocracy are ill-defined, every body lives in constant dread, lest advantage should be taken of his familiarity. Unable to judge, at once, of the social position of those he meets, an Englishman prudently[Pg 147] avoids all contact with them. Men are afraid, lest some slight service rendered should draw them into an unsuitable acquaintance; they dread civilities, and they avoid the obtrusive gratitude of a stranger, as much as his hatred."

Thus, *facts* seem to show that when the most aristocratic nation in the world is compared, as to manners, with the most democratic, the judgement of strangers is in favor of the latter.

And if good-manners are the outward exhibition of the democratic principle of impartial benevolence and equal rights, surely the nation which adopts this rule, both in social and civil life, is the most likely to secure the desirable exterior. The aristocrat, by his principles, extends the exterior of impartial benevolence to his own class, only; the democratic principle, requires it to be extended *to all*.

There is reason, therefore, to hope and expect more refined and polished manners in America, than in any other land; while all the developements of taste and refinement, such as poetry, music, painting, sculpture, and architecture, it may be expected, will come to a higher state of perfection, here, than in any other nation.

If this Country increases in virtue and intelligence, as it may, there is no end to the wealth which will pour in as the result of our resources of climate, soil, and navigation, and the skill, industry, energy, and enterprise, of our countrymen. This wealth, if used as intelligence and virtue dictate, will furnish the means for a superior education to all classes, and every facility for the refinement of taste, intellect, and feeling.

Moreover, in this Country, labor is ceasing to be the badge of a lower class; so that already it is disreputable for a man to be "a lazy gentleman." And this feeling must increase, till there is such an equalisation of labor, as will afford all the time needful for every class to improve the many advantages offered to them. Already, in Boston, through the munificence of some of her[Pg 148] citizens, there are literary and scientific advantages, offered to all classes, rarely enjoyed elsewhere. In Cincinnati, too, the advantages of education, now offered to the poorest classes, without charge, surpass what, some years ago, most wealthy men could purchase, for any price. And it is believed, that a time will come, when the poorest boy in America can secure advantages, which will equal what the heir of the proudest peerage can now command.

The records of the courts of France and Germany, (as detailed by the Duchess of Orleans,) in and succeeding the brilliant reign of Louis the Fourteenth,—a period which was deemed the acme of elegance and refinement,—exhibit a grossness, a vulgarity, and a coarseness, not to be found among the lowest of our respectable poor. And the biography of Beau Nash, who attempted to reform the manners of the gentry, in the times of Queen Anne, exhibits violations of the rules of decency among the aristocracy, which the commonest yeoman of this Land would feel disgraced in perpetrating.

This shows, that our lowest classes, at this period, are more refined, than were the highest in aristocratic lands, a hundred years ago; and another century may show the lowest classes, in wealth, in this Country, attaining as high a polish, as adorns those who now are leaders of good-manners in the courts of kings.

FOOTNOTE:

[M] The universal practice of this Nation, in thus giving precedence to woman, has been severely commented on by Miss Martineau and some others, who would transfer all the business of the other sex to women, and then have them treated like men. May this evidence of our superior civilisation and Christianity increase, rather than diminish!

CHAPTER XIII.
ON THE PRESERVATION OF A GOOD TEMPER IN A HOUSEKEEPER.

There is nothing, which has a more abiding influence on the happiness of a family, than the preservation of equable and cheerful temper and tones in the housekeeper. A woman, who is habitually gentle, sympathizing, [Pg 149]forbearing, and cheerful, carries an atmosphere about her, which imparts a soothing and sustaining influence, and renders it easier for all to do right, under her administration, than in any other situation.

The writer has known families, where the mother's presence seemed the sunshine of the circle around her; imparting a cheering and vivifying power, scarcely realized, till it was withdrawn. Every one, without thinking of it, or knowing why it was so, experienced a peaceful and invigorating influence, as soon as he entered the sphere illumined by her smile, and sustained by her cheering kindness and sympathy. On the contrary, many a good housekeeper, (good in every respect but this,) by wearing a countenance of anxiety and dissatisfaction, and by indulging in the frequent use of sharp and reprehensive tones, more than destroys all the comfort which otherwise would result from her system, neatness, and economy.

There is a secret, social sympathy, which every mind, to a greater or less degree, experiences with the feelings of those around, as they are manifested by the countenance and voice. A sorrowful, a discontented, or an angry, countenance, produces a silent, sympathetic influence, imparting a sombre shade to the mind, while tones of anger or complaint still more effectually jar the spirits.

No person can maintain a quiet and cheerful frame of mind, while tones of discontent and displeasure are sounding on the ear. We may gradually accustom ourselves to the evil, till it is partially diminished; but it always is an evil, which greatly interferes with the enjoyment of the family state. There are sometimes cases, where the entrance of the mistress of a family seems to awaken a slight apprehension, in every mind around, as if each felt in danger of a reproof, for something either perpetrated or neglected. A woman, who should go around her house with a small stinging snapper, which she habitually applied to those whom she met, would be encountered with feelings very much like[Pg 150] to those which are experienced by the inmates of a family, where the mistress often uses her countenance and voice, to inflict similar penalties for duties neglected.

Yet, there are many allowances to be made for housekeepers, who sometimes imperceptibly and unconsciously fall into such habits. A woman, who attempts to carry out any plans of system, order, and economy, and who has her feelings and habits conformed to certain rules, is constantly liable to have her plans crossed, and her taste violated, by the inexperience or inattention of those about her. And no housekeeper, whatever may be her habits, can escape the frequent recurrence of negligence or mistake, which interferes with her plans. It is probable, that there is no class of persons, in the world, who have such incessant trials of temper, and temptations to be fretful, as American housekeepers. For a housekeeper's business is not, like that of the other sex, limited to a particular department, for which previous preparation is made. It consists of ten thousand little disconnected items, which can never be so systematically arranged, that there is no daily jostling, somewhere. And in the best-regulated families, it is not unfrequently the case, that some act of forgetfulness or carelessness, from some member, will disarrange the business of the whole day, so that every hour will bring renewed occasion for annoyance. And the more strongly a woman realizes the value of time, and the importance of system and order, the more will she be tempted to irritability and complaint.

The following considerations, may aid in preparing a woman to meet such daily crosses, with even a cheerful temper and tones.

In the first place, a woman, who has charge of a large household, should regard her duties as dignified, important, and difficult. The mind is so made, as to be elevated and cheered by a sense of far-reaching influence and usefulness. A woman, who feels that she is a cipher, and that it makes little difference how she performs her duties, has far less to sustain and invigorate [Pg 151]her, than one, who truly estimates the importance of her station. A man, who feels that the destinies of a nation are turning on the judgement and skill with which he plans and executes, has a pressure of motive, and an elevation of feeling, which are great safeguards from all that is low, trivial, and degrading.

So, an American mother and housekeeper, who looks at her position in the aspect presented in the previous pages, and who rightly estimates the long train of influences which will pass down to thousands, whose destinies, from generation to generation, will be modified by those decisions of her will, which regulated the temper, principles, and habits, of her family, must be elevated above petty temptations, which would otherwise assail her.

Again, a housekeeper should feel that she really has great difficulties to meet and overcome. A person, who wrongly thinks there is little danger, can never maintain so faithful a guard, as one who rightly estimates the temptations which beset her. Nor can one, who thinks that they are trifling difficulties which she has to encounter, and trivial temptations, to which she must yield, so much enjoy the just reward of conscious virtue and self-control, as one who takes an opposite view of the subject.

A third method, is, for a woman deliberately to calculate on having her best-arranged plans interfered with, very often; and to be in such a state of preparation, that the evil will not come unawares. So complicated are the pursuits, and so diverse the habits of the various members of a family, that it is almost impossible for every one to avoid interfering with the plans and taste of a housekeeper, in some one point or another. It is, therefore, most wise, for a woman to keep the loins of her mind ever girt, to meet such collisions with a cheerful and quiet spirit.

Another important rule, is, to form all plans and arrangements in consistency with the means at command, and the character of those around. A woman, who has[Pg 152] a heedless husband, and young children, and incompetent domestics, ought not to make such plans, as one may properly form, who will not, in so many directions, meet embarrassment. She must aim at just so much as she can probably secure, and no more; and thus she will usually escape much temptation, and much of the irritation of disappointment.

The fifth, and a very important, consideration, is, that *system, economy,* and *neatness,* are valuable, only so far as they tend to promote the comfort and wellbeing of those affected. Some women seem to act under the impression, that these advantages *must* be secured, at all events, even if the comfort of the family be the sacrifice. True, it is very important that children grow up in habits of system, neatness, and order; and it is very desirable that the mother give them every incentive, both by precept and example: but it is still more important, that they grow up with amiable tempers, that they learn to meet the crosses of life with patience and cheerfulness; and nothing has a greater influence to secure this, than a mother's example. Whenever, therefore, a woman cannot accomplish her plans of neatness and order, without injury to her own temper, or to the temper of others, she ought to modify and reduce them, until she can.

The sixth method, relates to the government of the tones of voice. In many cases, when a woman's domestic arrangements are suddenly and seriously crossed, it is impossible not to feel some irritation. But it *is* always possible to refrain from angry tones. A woman can resolve, that, whatever happens, she will not speak, till she can do it in a calm and gentle manner. *Perfect silence* is a safe resort, when such control cannot be attained, as enables a person to speak calmly; and this determination, persevered in, will eventually be crowned with success.

Many persons seem to imagine, that tones of anger are needful, in order to secure prompt obedience. But observation has convinced the writer that they are *never*[Pg 153] necessary; that *in all cases*, reproof, administered in calm tones, would be better. A case will be given in illustration.

A young girl had been repeatedly charged to avoid a certain arrangement in cooking. On one day, when company was invited to dine, the direction was forgotten, and the consequence was, an accident, which disarranged every thing, seriously injured the principal dish, and delayed dinner for an hour. The mistress of the family entered the kitchen, just as it occurred, and, at a glance, saw the extent of the mischief. For a moment, her eyes flashed, and her cheeks glowed; but she held her peace. After a minute or so, she gave directions, in a calm voice, as to the best mode of retrieving the evil, and then left, without a word said to the offender.

After the company left, she sent for the girl, alone, and in a calm and kind manner pointed out the aggravations of the case, and described the trouble which had been caused to her husband, her visiters, and herself. She then portrayed the future evils which would result from such habits of neglect and inattention, and the modes of attempting to overcome them; and then offered a reward for the future, if, in a given time, she succeeded in improving in this respect. Not a tone of anger was uttered; and yet the severest scolding of a practised Xantippe could not have secured such contrition, and determination to reform, as was gained by this method.

But similar negligence is often visited by a continuous stream of complaint and reproof, which, in most cases, is met, either by sullen silence, or impertinent retort, while anger prevents any contrition, or any resolution of future amendment.

It is very certain, that some ladies do carry forward a most efficient government, both of children and domestics, without employing tones of anger; and therefore they are not indispensable, nor on any account desirable.

[Pg 154]

Though some ladies, of intelligence and refinement, do fall unconsciously into such a practice, it is certainly very unlady-like, and in very bad taste, to *scold*; and the further a woman departs from all approach to it, the more perfectly she sustains her character as a lady.

Another method of securing equanimity, amid the trials of domestic life, is, to cultivate a habit of making allowances for the difficulties, ignorance, or temptations, of those who violate rule or neglect duty. It is vain, and most unreasonable, to expect the consideration and care of a mature mind, in childhood and youth; or that persons, of such limited advantages as most domestics have enjoyed, should practise proper self-control, and possess proper habits and principles.

Every parent, and every employer, needs daily to cultivate the spirit expressed in the Divine prayer, "forgive us our trespasses, as we forgive those who trespass against us." The same allowances and forbearance, which we supplicate from our Heavenly Father, and desire from our fellow-men, in reference to our own deficiencies, we should constantly aim to extend to all, who cross our feelings and interfere with our plans.

The last, and most important, mode of securing a placid and cheerful temper and tones, is, by a right view of the doctrine of a superintending Providence. All persons are too much in the habit of regarding the more important events of life, as exclusively under the control of Perfect Wisdom. But the fall of a sparrow, or the loss of a hair, they do not feel to be equally the result of His directing agency. In consequence of this, Christian persons, who aim at perfect and cheerful submission to heavy afflictions, and who succeed, to the edification of all about them, are sometimes sadly deficient under petty crosses. If a beloved child be laid in the grave, even if its death resulted from the carelessness of a domestic, or of a physician, the eye is turned from the subordinate agent, to the Supreme Guardian of all, and to Him they bow, without murmur or complaint. But if a pudding be burnt, or a[Pg 155] room badly swept, or an errand forgotten, then vexation and complaint are allowed, just as if these events were not appointed by Perfect Wisdom, as much as the sorer chastisement.

A woman, therefore, needs to cultivate the *habitual* feeling, that all the events of her nursery and kitchen, are brought about by the permission of our Heavenly Father, and that fretfulness or complaint, in regard to these, is, in fact, complaining and disputing at the appointments of God, and is really as sinful, as unsubmissive murmurs amid the sorer chastisements of His hand. And a woman, who cultivates this habit of referring all the minor trials of life to the wise and benevolent agency of a Heavenly Parent, and daily seeks His sympathy and aid, to enable her to meet them with a quiet and cheerful spirit, will soon find it the perennial spring of abiding peace and content.

CHAPTER XIV.
ON HABITS OF SYSTEM AND ORDER.

The discussion of the question of the equality of the sexes, in intellectual capacity, seems frivolous and useless, both because it can never be decided, and because there would be no possible advantage in the decision. But one topic, which is often drawn into this discussion, is of far more consequence; and that is, the relative importance and difficulty of the duties a woman is called to perform.

It is generally assumed, and almost as generally conceded, that woman's business and cares are contracted and trivial; and that the proper discharge of her duties, demands far less expansion of mind and vigor of intellect, than the pursuits of the other sex. This idea has prevailed, because women, as a mass, have never been educated with reference to their most important duties;[Pg 156] while that portion of their employments, which is of least value, has been regarded as the chief, if not the sole, concern of a woman. The covering of the body, the conveniences of residences, and the gratification of the appetite, have been too much regarded as the sole objects, on which her intellectual powers are to be exercised.

But, as society gradually shakes off the remnants of barbarism, and the intellectual and moral interests of man rise, in estimation, above the merely sensual, a truer estimate is formed of woman's duties, and of the measure of intellect requisite for the proper discharge of them. Let any man, of sense and discernment, become the member of a large household, in which, a well-educated and pious woman is endeavoring systematically to discharge her multiform duties; let him fully comprehend all her cares, difficulties, and perplexities; and it is probable he would coincide in the opinion, that no statesman, at the head of a nation's affairs, had more frequent calls for wisdom, firmness, tact, discrimination, prudence, and versatility of talent, than such a woman.

She has a husband, to whose peculiar tastes and habits she must accommodate herself; she has children, whose health she must guard, whose physical constitutions she must study and develope, whose temper and habits she must regulate, whose principles she must form, whose pursuits she must direct. She has constantly changing domestics, with all varieties of temper and habits, whom she must govern, instruct, and direct; she is required to regulate the finances of the domestic state, and constantly to adapt expenditures to the means and to the relative claims of each department. She has the direction of the kitchen, where ignorance, forgetfulness, and awkwardness, are to be so regulated, that the various operations shall each start at the right time, and all be in completeness at the same given hour. She has the claims of society to meet, calls to receive and return, and the duties of hospitality to sustain. She has the poor to relieve; benevolent societies to aid;[Pg 157] the schools of her children to inquire and decide about; the care of the sick; the nursing of infancy; and the endless miscellany of odd items, constantly recurring in a large family.

Surely, it is a pernicious and mistaken idea, that the duties, which tax a woman's mind, are petty, trivial, or unworthy of the highest grade of intellect and moral worth. Instead of allowing this feeling, every woman should imbibe, from early youth, the impression, that she is training for the discharge of the most important, the most difficult, and the most sacred and interesting duties that can possibly employ the highest intellect. She ought to feel, that her station and responsibilities, in the great drama of life, are second to none, either as viewed by her Maker, or in the estimation of all minds whose judgement is most worthy of respect.

She, who is the mother and housekeeper in a large family, is the sovereign of an empire, demanding more varied cares, and involving more difficult duties, than are really exacted of her, who, while she wears the crown, and professedly regulates the interests of the greatest nation on earth, finds abundant leisure for theatres, balls, horseraces, and every gay pursuit.

There is no one thing, more necessary to a housekeeper, in performing her varied duties, than *a habit of system and order*; and yet, the peculiarly desultory nature of women's pursuits, and the embarrassments resulting from the state of domestic service in this Country, render it very difficult to form such a habit. But it is sometimes the case, that women, who could and would carry forward a systematic plan of domestic economy, do not attempt it, simply from a want of knowledge of the various modes of introducing it. It is with reference to such, that various modes of securing system and order, which the writer has seen adopted, will be pointed out.

A wise economy is nowhere more conspicuous, than in the right *apportionment of time* to different pursuits. There are duties of a religious, intellectual, social, and[Pg 158] domestic, nature, each having different relative claims on attention. Unless a person has some general plan of apportioning these claims, some will intrench on others, and some, it is probable, will be entirely excluded. Thus, some find religious, social, and domestic, duties, so numerous, that no time is given to intellectual improvement. Others, find either social, or benevolent, or religious, interests, excluded by the extent and variety of other engagements.

It is wise, therefore, for all persons to devise a general plan, which they will at least keep in view, and aim to accomplish, and by which, a proper proportion of time shall be secured, for all the duties of life.

In forming such a plan, every woman must accommodate herself to the peculiarities of her situation. If she has a large family, and a small income, she must devote far more time to the simple duty of providing food and raiment, than would be right were she in affluence, and with a small family. It is impossible, therefore, to draw out any general plan, which all can adopt. But there are some *general principles*, which ought to be the guiding rules, when a woman arranges her domestic employments. These principles are to be based on Christianity, which teaches us to "seek first the kingdom of God," and to deem food, raiment, and the conveniences of life, as of secondary account. Every woman, then, ought to start with the assumption, that religion is of more consequence than any worldly concern, and that, whatever else may be sacrificed, this, shall be the leading object, in all her arrangements, in respect to time, money, and attention. It is also one of the plainest requisitions of Christianity, that we devote some of our time and efforts, to the comfort and improvement of others. There is no duty, so constantly enforced, both in the Old and New Testament, as the duty of charity, in dispensing to those, who are destitute of the blessings we enjoy. In selecting objects of charity, the same rule applies to others, as to ourselves; their moral and religious interests are of the[Pg 159] highest moment, and for them, as well as for ourselves, we are to "seek first the kingdom of God."

Another general principle, is, that our intellectual and social interests are to be preferred, to the mere gratification of taste or appetite. A portion of time, therefore, must be devoted to the cultivation of the intellect and the social affections.

Another, is, that the mere gratification of appetite, is to be placed *last* in our estimate; so that, when a question arises, as to which shall be sacrificed, some intellectual, moral, or social, advantage, or some gratification of sense, we should invariably sacrifice the last.

Another, is, that, as health is indispensable to the discharge of every duty, nothing, which sacrifices that blessing, is to be allowed, in order to gain any other advantage or enjoyment. There are emergencies, when it is right to risk health and life, to save ourselves and others from greater evils; but these are exceptions, which do not militate against the general rule. Many persons imagine, that, if they violate the laws of health, in performing religious or domestic duties, they are guiltless before God. But such greatly mistake. We as directly violate the law, "thou shalt not kill," when we do what tends to risk or shorten our own life, as if we should intentionally run a dagger into a neighbor. True, we may escape any fatal or permanently injurious effects, and so may a dagger or bullet miss the mark, or do only transient injury. But this, in either case, makes the sin none the less. The life and happiness of all His creatures are dear to our Creator; and He is as much displeased, when we injure our own interests, as when we injure those of others. The idea, therefore, that we are excusable, if we harm no one but ourselves, is false and pernicious. These, then, are the general principles, to guide a woman in systematizing her duties and pursuits.

The Creator of all things, is a Being of perfect system and order; and, to aid us in our duty, in this respect, He has divided our time, by a regularly returning [Pg 160]day of rest from worldly business. In following this example, the intervening six days may be subdivided to secure similar benefits. In doing this, a certain portion of time must be given to procure the means of livelihood, and for preparing food, raiment, and dwellings. To these objects, some must devote more, and others less, attention. The remainder of time not necessarily thus employed, might be divided somewhat in this manner: The leisure of two afternoons and evenings, could be devoted to religious and benevolent objects, such as religious meetings, charitable associations, school visiting, and attention to the sick and poor. The leisure of two other days, might be devoted to intellectual improvement, and the pursuits of taste. The leisure of another day, might be devoted to social enjoyments, in making or receiving visits; and that of another, to miscellaneous domestic pursuits, not included in the other particulars.

It is probable, that few persons could carry out such an arrangement, very strictly; but every one can make a systematic apportionment of time, and at least *aim* at accomplishing it; and they can also compare the time which they actually devote to these different objects, with such a general outline, for the purpose of modifying any mistaken proportions.

Without attempting any such systematic employment of time, and carrying it out, so far as they can control circumstances, most women are rather driven along, by the daily occurrences of life, so that, instead of being the intelligent regulators of their own time, they are the mere sport of circumstances. There is nothing, which so distinctly marks the difference between weak and strong minds, as the fact, whether they control circumstances, or circumstances control them.

It is very much to be feared, that the apportionment of time, actually made by most women, exactly inverts the order, required by reason and Christianity. Thus, the furnishing a needless variety of food, the conveniences of dwellings, and the adornments of dress,[Pg 161] often take a larger portion of time, than is given to any other object. Next after this, comes intellectual improvement; and, last of all, benevolence and religion.

It may be urged, that it is indispensable for most persons to give more time to earn a livelihood, and to prepare food, raiment, and dwellings, than to any other object. But it may be asked, how much of the time, devoted to these objects, is employed in preparing varieties of food, not necessary, but rather injurious, and how much is spent for those parts of dress and furniture not indispensable, and merely ornamental? Let a woman subtract from her domestic employments, all the time, given to pursuits which are of no use, except as they gratify a taste for ornament, or minister increased varieties, to tempt the appetite, and she will find, that much, which she calls "domestic duties," and which prevent her attention to intellectual, benevolent, and religious, objects, should be called by a very different name. No woman has a right to give up attention to the higher interests of herself and others, for the ornaments of taste, or the gratification of the palate. To a certain extent, these lower objects are lawful and desirable; but, when they intrude on nobler interests, they become selfish and degrading. Every woman, then, when employing her hands, in ornamenting her person, her children, or her house, ought to calculate, whether she has devoted *as much* time, to the intellectual and moral wants of herself and others. If she has not, she may know that she is doing wrong, and that her system, for apportioning her time and pursuits, should be altered.

Some persons, endeavor to systematize their pursuits, by apportioning them to particular hours of each day. For example, a certain period before breakfast, is given to devotional duties; after breakfast, certain hours are devoted to exercise and domestic employments; other hours, to sewing, or reading, or visiting; and others, to benevolent duties. But, in most cases, it is more[Pg 162] difficult to systematize the hours of each day, than it is to secure some regular division of the week.

In regard to the minutiæ of domestic arrangements, the writer has known the following methods to be adopted. *Monday*, with some of the best housekeepers, is devoted to preparing for the labors of the week. Any extra cooking, the purchasing of articles to be used during the week, the assorting of clothes for the wash, and mending such as would be injured without;—these, and similar items, belong to this day. *Tuesday* is devoted to washing, and *Wednesday* to ironing. On *Thursday*, the ironing is finished off, the clothes are folded and put away, and all articles, which need mending, are put in the mending basket, and attended to. *Friday* is devoted to sweeping and housecleaning. On *Saturday*, and especially the last Saturday of every month, every department is put in order; the castors and table furniture are regulated, the pantry and cellar inspected, the trunks, drawers, and closets arranged, and every thing about the house, put in order for *Sunday*. All the cooking, needed for Sunday, is also prepared. By this regular recurrence of a particular time, for inspecting every thing, nothing is forgotten till ruined by neglect.

Another mode of systematizing, relates to providing proper supplies of conveniences, and proper places in which to keep them. Thus, some ladies keep a large closet, in which are placed the tubs, pails, dippers, soap-dishes, starch, bluing, clothes-line, clothes-pins, and every other article used in washing; and in the same, or another, place, are kept every convenience for ironing. In the sewing department, a trunk, with suitable partitions, is provided, in which are placed, each in its proper place, white thread of all sizes, colored thread, yarns for mending, colored and black sewing-silks and twist, tapes and bobbins of all sizes, white and colored welting-cords, silk braids and cords, needles of all sizes, papers of pins, remnants of linen and colored cambric, a supply of all kinds of buttons used[Pg 163] in the family, black and white hooks and eyes, a yard measure, and all the patterns used in cutting and fitting. These are done up in separate parcels, and labelled. In another trunk, are kept all pieces used in mending, arranged in order, so that any article can be found, without loss of time. A trunk, like the first mentioned, will save many steps, and often much time and perplexity; while by purchasing articles thus by the quantity, they come much cheaper, than if bought in little portions as they are wanted. Such a trunk should be kept locked, and a smaller supply, for current use, retained in a workbasket.

A full supply of all conveniences in the kitchen and cellar, and a place appointed for each article, very much facilitates domestic labor. For want of this, much vexation and loss of time is occasioned, while seeking vessels in use, or in cleansing those employed by different persons, for various purposes. It would be far better, for a lady to give up some expensive article, in the parlor, and apply the money, thus saved, for kitchen conveniences, than to have a stinted supply, where the most labor is to be performed. If our Countrywomen would devote more to comfort and convenience, and less to show, it would be a great improvement. Expensive mirrors and pier-tables in the parlor, and an unpainted, gloomy, ill-furnished kitchen, not unfrequently are found under the same roof.

Another important item, in systematic economy, is, the apportioning of *regular* employment to the various members of a family. If a housekeeper can secure the cooperation of *all* her family, she will find, that "many hands make light work." There is no greater mistake, than in bringing up children to feel that they must be taken care of, and waited on, by others, without any corresponding obligations on their part. The extent, to which young children can be made useful, in a family, would seem surprising, to those who have never seen a *systematic* and *regular* plan for securing[Pg 164] their services. The writer has been in a family, where a little girl, of eight or nine years of age, washed and dressed herself and young brother, and made their small beds, before breakfast, set and cleared all the tables, at meals, with a little help from a grown person in moving tables and spreading cloths, while all the dusting of parlors and chambers was also neatly performed by her. A brother, of ten years old, brought in and piled all the wood, used in the kitchen and parlor, brushed the boots and shoes, neatly, went on errands, and took all the care of the poultry. They were children, whose parents could afford to hire servants to do this, but who chose to have their children grow up healthy and industrious, while proper instruction, system, and encouragement, made these services rather a pleasure, than otherwise, to the children.

Some parents pay their children for such services; but this is hazardous, as tending to make them feel that they are not bound to be helpful without pay, and also as tending to produce a hoarding, money-making spirit. But, where children have no hoarding propensities, and need to acquire a sense of the value of property, it may be well to let them earn money, for some extra services, rather as a favor. When this is done, they should be taught to spend it for others, as well as for themselves; and in this way, a generous and liberal spirit will be cultivated.

There are some mothers, who take pains to teach their boys most of the domestic arts, which their sisters learn. The writer has seen boys, mending their own garments, and aiding their mother or sisters in the kitchen, with great skill and adroitness; and at an early age, they usually very much relish joining in such occupations. The sons of such mothers, in their college life, or in roaming about the world, or in nursing a sick wife or infant, find occasion to bless the forethought and kindness, which prepared them for such emergencies. Few things are in worse taste, than for a man needlessly to busy himself in women's work;[Pg 165] and yet a man never appears in a more interesting attitude, than when, by skill in such matters, he can save a mother or wife from care and suffering. The more a boy is taught to use his hands, in every variety of domestic employment, the more his faculties, both of mind and body, are developed; for mechanical pursuits exercise the intellect, as well as the hands. The early training of New-England boys, in which they turn their hand to almost every thing, is one great reason of the quick perceptions, versatility of mind, and mechanical skill, for which that portion of our Countrymen is distinguished.

The writer has known one mode of systematizing the aid of the older children in a family, which, in some cases of very large families, it may be well to imitate. In the case referred to, when the oldest daughter was eight or nine years old, an infant sister was given to her, as her special charge. She tended it, made and mended its clothes, taught it to read, and was its nurse and guardian, through all its childhood. Another infant was given to the next daughter, and thus the children were all paired in this interesting relation. In addition to the relief thus afforded to the mother, the elder children were in this way qualified for their future domestic relations, and both older and younger bound to each other by peculiar ties of tenderness and gratitude.

In offering these examples, of various modes of systematizing, one suggestion may be worthy of attention. It is not unfrequently the case, that ladies, who find themselves cumbered with oppressive cares, after reading remarks on the benefits of system, immediately commence the task of arranging their pursuits, with great vigor and hope. They divide the day into regular periods, and give each hour its duty; they systematize their work, and endeavor to bring every thing into a regular routine. But, in a short time, they find themselves baffled, discouraged, and disheartened, and finally relapse into their former desultory ways, in[Pg 166] a sort of resigned despair. The difficulty, in such cases, is, that they attempt too much at a time. There is nothing, which so much depends upon *habit*, as a systematic mode of performing duty; and, where no such habit has been formed, it is impossible for a novice to start, at once, into a universal mode of systematizing, which none but an adept could carry through. The only way for such persons, is, to begin with a little at a time. Let them select some three or four things, and resolutely attempt to conquer at these points. In time, a habit will be formed, of doing a few things at regular periods, and in a systematic way. Then it will be easy to add a few more; and thus, by a gradual process, the object can be secured, which it would be vain to attempt, by a more summary course. Early rising is almost an indispensable condition to success, in such an effort; but, where a woman lacks either the health or the energy to secure a period for devotional duties before breakfast, let her select that hour of the day, in which she will be least liable to interruption, and let her then seek strength and wisdom from the only true Source. At this time, let her take a pen, and make a list of all the things which she considers as duties. Then, let a calculation be made, whether there be time enough, in the day or the week, for all these duties. If there be not, let the least important be stricken from the list, as not being duties, and which must be omitted. In doing this, let a woman remember, that, though "what we shall eat, and what we shall drink, and wherewithal we shall be clothed," are matters requiring due attention, they are very apt to obtain a wrong relative importance, while social, intellectual, and moral, interests, receive too little regard.

In this Country, eating, dressing, and household furniture and ornaments, take far too large a place in the estimate of relative importance; and it is probable, that most women could modify their views and practice, so as to come nearer to the Saviour's requirements. No woman has a right to put a stitch of ornament [Pg 167]on any article of dress or furniture, or to provide one superfluity in food, until she is sure she can secure time for all her social, intellectual, benevolent, and religious, duties. If a woman will take the trouble to make such a calculation as this, she will usually find that she has time enough, to perform all her duties easily and well.

It is impossible, for a conscientious woman to secure that peaceful mind, and cheerful enjoyment of life, which all should seek, who is constantly finding her duties jarring with each other, and much remaining undone, which she feels that she ought to do. In consequence of this, there will be a secret uneasiness, which will throw a shade over the whole current of life, never to be removed, till she so efficiently defines and regulates her duties, that she can fulfil them all.

And here the writer would urge upon young ladies, the importance of forming habits of system, while unembarrassed with those multiplied cares, which will make the task so much more difficult and hopeless. Every young lady can systematize her pursuits, to a certain extent. She can have a particular day for mending her wardrobe, and for arranging her trunks, closets, and drawers. She can keep her workbasket, her desk at school, and all her other conveniences, in their proper places, and in regular order. She can have regular periods for reading, walking, visiting, study, and domestic pursuits. And, by following this method, in youth, she will form a taste for regularity, and a habit of system, which will prove a blessing to her, through life.

CHAPTER XV.
ON GIVING IN CHARITY.

It is probable, that there is no point of duty, where conscientious persons differ more in opinion, or where they find it more difficult to form discriminating and[Pg 168] decided views, than on the matter of charity. That we are bound to give *some* of our time, money, and efforts, to relieve the destitute, all allow. But, as to how much we are to give, and on whom our charities shall be bestowed, many a reflecting mind has been at a loss. Yet it seems very desirable, that, in reference to a duty so constantly and so strenuously urged by the Supreme Ruler, we should be able so to fix metes and bounds, as to keep a conscience void of offence, and to free the mind from disquieting fears of deficiency.

The writer has found no other topic of investigation so beset with difficulty, and so absolutely without the range of definite rules, which can apply to all, in all circumstances. But on this, as on a previous topic, there seem to be *general principles*, by the aid of which, any candid mind, sincerely desirous of obeying the commands of Christ, however much self-denial may be involved, can arrive at definite conclusions, as to its own individual obligations, so that, when these are fulfilled, the mind may be at peace.

But, for a mind that is worldly, living mainly to seek its own pleasures, instead of living to please God, no principles can be so fixed, as not to leave a ready escape from all obligation. Such minds, either by indolence (and consequent ignorance) or by sophistry, will convince themselves, that a life of engrossing self-indulgence, with perhaps the gift of a few dollars, and a few hours of time, may suffice, to fulfil the requisitions of the Eternal Judge.

For such minds, no reasonings will avail, till the heart is so changed, that, to learn the will and follow the example of Jesus Christ, become the leading objects of interest and effort. It is to aid those, who profess to possess this temper of mind, that the following suggestions are offered.

The first consideration, which gives definiteness to this subject, is, a correct view of the object for which we are placed in this world. A great many even of professed Christians, seem to be acting on the supposition, that the object of life is to secure as much as[Pg 169] possible of all the various enjoyments placed within reach. Not so, teaches reason or revelation. From these, we learn, that, though the happiness of His creatures, is the end for which God created and sustains them, yet, that this happiness depends, not on the various modes of gratification put within our reach, but mainly on *character*. A man may possess all the resources for enjoyment which this world can afford, and yet feel that "all is vanity and vexation of spirit," and that he is supremely wretched. Another, may be in want of all things, and yet possess that living spring of benevolence, faith, and hope, which will make an Eden of the darkest prison.

In order to be perfectly happy, man must attain that character, which Christ exhibited; and the nearer he approaches it, the more will happiness reign in his breast.

But what was the grand peculiarity of the character of Christ? It was *self-denying benevolence*. He came not to "seek His own;" He "went about doing good," and this was His "meat and drink;" that is, it was this which sustained the health and life of His mind, as food and drink sustain the health and life of the body. Now, the mind of man is so made, that it can gradually be transformed into the same likeness. A selfish being, who, for a whole life, has been nourishing habits of indolent self-indulgence, can, by taking Christ as his example, by communion with Him, and by daily striving to imitate His character and conduct, form such a temper of mind, that "doing good" will become the chief and highest source of enjoyment. And this heavenly principle will grow stronger and stronger, until self-denial loses the more painful part of its character, and then, *living to make happiness*, will be so delightful and absorbing a pursuit, that all exertions, regarded as the means to this end, will be like the joyous efforts of men, when they strive for a prize or a crown, with the full hope of success.

In this view of the subject, efforts and self-denial, for the good of others, are to be regarded, not merely as[Pg 170] duties enjoined for the benefit of others, but as the moral training indispensable to the formation of that character, on which depends our own happiness. This view, exhibits the full meaning of the Saviour's declaration, "how hardly shall they that have riches enter into the kingdom of God!" He had before taught, that the kingdom of Heaven consisted, not in such enjoyments as the worldly seek, but, in the temper of self-denying benevolence, like His own; and, as the rich have far greater temptations to indolent self-indulgence, they are far less likely to acquire this temper, than those, who, by limited means, are inured to some degree of self-denial.

But, on this point, one important distinction needs to be made; and that is, between the self-denial, which has no other aim than mere self-mortification, and that, which is exercised to secure greater good to ourselves and others. The first is the foundation of monasticism, penances, and all other forms of asceticism; the latter, only, is that which Christianity requires.

A second consideration, which may give definiteness to this subject, is, that the formation of a perfect character, involves, not the extermination of any principles of our nature, but rather the regulating of them, according to the rules of reason and religion; so that the lower propensities shall always be kept subordinate to nobler principles. Thus, we are not to aim at destroying our appetites, or at needlessly denying them, but rather so to regulate them, that they shall best secure the objects for which they were implanted. We are not to annihilate the love of praise and admiration; but so to control it, that the favor of God shall be regarded more than the estimation of men. We are not to extirpate the principle of curiosity, which leads us to acquire knowledge; but so to direct it, that all our acquisitions shall be useful and not frivolous or injurious. And thus, with all the principles of the mind, God has implanted no desires in our constitution, which are evil and pernicious. On the contrary, all our constitutional[Pg 171] propensities, either of mind or body, He designed we should gratify, whenever no evils would thence result, either to ourselves or others. Such passions as envy, ambition, pride, revenge, and hatred, are to be exterminated; for they are either excesses or excrescences: not created by God, but rather the result of our own neglect to form habits of benevolence and self-control.

In deciding the rules of our conduct, therefore, we are ever to bear in mind, that the developement of the nobler principles, and the subjugation of inferior propensities to them, is to be the main object of effort, both for ourselves and for others. And, in conformity with this, in all our plans, we are to place religious and moral interests as first in estimation, our social and intellectual interests, next, and our physical gratifications, as subordinate to all.

A third consideration, is, that, though the means for sustaining life and health are to be regarded as necessaries, without which no other duties can be performed, yet, that a very large portion of the time, spent by most persons, in easy circumstances, for food, raiment, and dwellings, are for mere *superfluities*, which *are right, when they do not involve the sacrifice of higher interests*, and *wrong, when they do*. Life and health can be sustained in the humblest dwellings, with the plainest dress, and the simplest food; and, after taking from our means, what is necessary for life and health, the remainder is to be so divided, that the larger portion shall be given to supply the moral and intellectual wants of ourselves and others, and the smaller share to procure those additional gratifications, of taste and appetite, which are desirable, but not indispensable. Mankind, thus far, have never made this apportionment of their means; yet, just as fast as they have risen from a savage state, mere physical wants have been made, to an increasing extent, subordinate to higher objects.

Another very important consideration, is, that, in urging the duty of charity, and the prior claims of moral and religious objects, no rule of duty should be[Pg 172] maintained, which it would not be right and wise for *all* to follow. And we are to test the wisdom of any general rule, by inquiring what would be the result, if all mankind should practise according to it. In view of this, we are enabled to judge of the correctness of those, who maintain, that, to be consistent, men believing in the eternal destruction of all those of our race who are not brought under the influence of the Christian system, should give up, not merely the elegances, but all the superfluities, of life, and devote the whole of their means, not indispensable to life and health, for the propagation of Christianity. But, if this is the duty of any, it is the duty of all; and we are to inquire what would be the result, if all conscientious persons gave up the use of all superfluities. Suppose, that two millions of the people in the United States, were conscientious persons, and relinquished the use of every thing not absolutely necessary to life and health. It would instantly throw out of employment one half of the whole community. The manufacturers, mechanics, merchants, agriculturists, and all the agencies they employ, would be beggared, and one half of those not reduced to poverty, would be obliged to spend all their extra means, in simply supplying necessaries to the other half. The use of superfluities, therefore, to a certain extent, is as indispensable to promote industry, virtue, and religion, as any direct giving of money or time; and it is owing entirely to a want of reflection, and of comprehensive views, that any men ever make so great a mistake, as is here exhibited.

Instead, then, of urging a rule of duty which is at once irrational and impracticable, there is another course, which commends itself to the understandings of all. For whatever may be the *practice*, of intelligent men, they universally concede the *principle*, that our physical gratifications should always be made subordinate to social, intellectual, and moral, advantages. And all that is required, for the advancement of our whole race to the most perfect state of society, is, simply, that[Pg 173] men should act in agreement with this principle. And, if only a very small portion, of the most intelligent of our race, should act according to this rule, under the control of Christian benevolence, the immense supplies, furnished, for the general good, would be far beyond what any would imagine, who had never made any calculations on the subject. In this Nation, alone, suppose the one million and more, of professed followers of Christ, should give a larger portion of their means, for the social, intellectual, and moral, wants of mankind, than for the superfluities that minister to taste, convenience, and appetite; it would be enough to furnish all the schools, colleges, Bibles, ministers, and missionaries, that the whole world could demand; or, at least, it would be far more, than properly qualified agents to administer it, could employ.

But, it may be objected, that, though this view is one, which, in the abstract, looks plausible and rational, not one in a thousand, can practically adopt it. How few keep any account, at all, of their current expenses! How impossible it is, to determine, exactly, what are necessaries, and what are superfluities! And in regard to women, how few have the control of an income, so as not to be bound by the wishes of a parent or a husband!

In reference to these difficulties, the first remark is, that we are never under obligations to do, what is entirely out of our power, so that those persons, who have no power to regulate their expenses or their charities, are under no sort of obligation to attempt it. The second remark is, that, when a rule of duty is discovered, we are bound to *aim* at it, and to fulfil it, just so far as we can. We have no right to throw it aside, because we shall find some difficult cases, when we come to apply it. The third remark is, that no person can tell how much can be done, till a faithful trial has been made. If a woman has never kept any accounts, nor attempted to regulate her expenditures by the right rule, nor used her influence with those that control her[Pg 174] plans, to secure this object, she has no right to say how much she can, or cannot, do, till after a fair trial has been made.

In attempting such a trial, the following method can be taken. Let a woman keep an account of all she spends, for herself and her family, for a year, arranging the items under three general heads. Under the first, put all articles for food, raiment, rent, wages, and all conveniences. Under the second, place all sums paid in securing an education, and books, and other intellectual advantages. Under the third head, place all that is spent for benevolence and religion. At the end of the year, the first and largest account will show the mixed items of necessaries and superfluities, which can be arranged, so as to gain some sort of idea how much has been spent for superfluities, and how much for necessaries. Then, by comparing what is spent for superfluities, with what is spent for intellectual and moral advantages, data will be gained, for judging of the past, and regulating the future.

Does a woman say she cannot do this? let her inquire, whether the offer of a thousand dollars, as a reward for attempting it one year, would not make her undertake to do it; and, if so, let her decide, in her own mind, which is most valuable, a clear conscience, and the approbation of God, in this effort to do His will, or one thousand dollars. And let her do it, with this warning of the Saviour before her eyes, — "No man can serve two masters." "Ye cannot serve God and Mammon."

Is it objected, How can we decide between superfluities and necessaries, in this list? it is replied, that we are not required to judge exactly, in all cases. Our duty is, to use the means in our power to assist us in forming a correct judgement; to seek the Divine aid in freeing our minds from indolence and selfishness; and then to judge as well as we can, in our endeavors rightly to apportion and regulate our expenses. Many persons seem to feel that they are bound to do better[Pg 175] than they know how. But God is not so hard a Master; and, after we have used all proper means to learn the right way, if we then follow it, according to our ability, we do wrong to feel misgivings, or to blame ourselves, if results come out differently from what seems desirable. The results of our actions, alone, can never prove us deserving of blame. For men are often so placed, that, owing to lack of intellect or means, it is impossible for them to decide correctly. To use all the means of knowledge within our reach, and then to judge, with a candid and conscientious spirit, is all that God requires; and, when we have done this, and the event seems to come out wrong, we should never wish that we had decided otherwise. For it is the same as wishing that we had not followed the dictates of judgement and conscience. As this is a world designed for discipline and trial, untoward events are never to be construed as indications of the obliquity of our past decisions.

But it is probable, that a great portion of the women of this Nation, cannot secure any such systematic mode of regulating their expenses. To such, the writer would propose one inquiry; cannot you calculate how much *time* and *money* you spend for what is merely ornamental, and not necessary, for yourself, your children, and your house? Cannot you compare this with the time and money you spend for intellectual and benevolent purposes? and will not this show the need of some change? In making this examination, is not this brief rule, deducible from the principles before laid down, the one which should regulate you? Every person does right, in spending *some* portion of time and means in securing the conveniences and adornments of taste; but the amount should never exceed what is spent in securing our own moral and intellectual improvement, nor exceed what is spent in benevolent efforts to supply the physical and moral wants of our fellow-men.

In making an examination on this subject, it is sometimes the case, that a woman will count among the[Pg 176]*necessaries* of life, all the various modes of adorning the person or house, practised in the circle in which she moves; and, after enumerating the many *duties* which demand attention, counting these as a part, she will come to the conclusion, that she has no time, and but little money, to devote to personal improvement, or to benevolent enterprises. This surely is not in agreement with the requirements of the Saviour, who calls on us to seek for others, as well as ourselves, *first of all*, "the kingdom of God, and His righteousness."

In order to act in accordance with the rule here presented, it is true, that many would be obliged to give up the idea of conforming to the notions and customs of those, with whom they associate, and compelled to adopt the maxim, "be not conformed to this world." In many cases, it would involve an entire change in the style of living. And the writer has the happiness of knowing more cases than one, where persons, who have come to similar views, on this subject, have given up large and expensive establishments, disposed of their carriages, dismissed a portion of their domestics, and modified all their expenditures, that they might keep a pure conscience, and regulate their charities more according to the requirements of Christianity. And there are persons, well known in the religious world, who save themselves all labor of minute calculation, by devoting so large a portion of their time and means to benevolent objects, that they find no difficulty in knowing that they give more for religious, benevolent, and intellectual, purposes, than for superfluities.

In deciding what particular objects shall receive our benefactions, there are also general principles to guide us. The first, is that presented by our Saviour, when, after urging the great law of benevolence, He was asked, "and who is my neighbor?" His reply, in the parable of 'the Good Samaritan,' teaches us, that any human being, whose wants are brought to our knowledge, is our neighbor. The wounded man was not only a stranger, but he belonged to a foreign nation,[Pg 177] peculiarly hated; and he had no claim, except that his wants were brought to the knowledge of the wayfaring man. From this, we learn, that the destitute, of all nations, become our neighbors, as soon as their wants are brought to our knowledge.

Another general principle, is this, that those who are most in need, must be relieved, in preference to those who are less destitute. On this principle, it is, that we think the followers of Christ should give more to supply those who are suffering for want of the bread of eternal life, than for those who are deprived of physical enjoyments. And another reason for this preference, is, the fact, that many, who give in charity, have made such imperfect advances in civilization and Christianity, that the intellectual and moral wants of our race make but a feeble impression on the mind. Relate a pitiful tale of a family, reduced to live, for weeks, on potatoes, only, and many a mind would awake to deep sympathy, and stretch forth the hand of charity. But describe cases, where the immortal mind is pining in stupidity and ignorance, or racked with the fever of baleful passions, and how small the number, so elevated in sentiment, and so enlarged in their views, as to appreciate and sympathize in these far greater misfortunes! The intellectual and moral wants of our fellow-men, therefore, should claim the first place in our attention, both because they are most important, and because they are most neglected.

Another consideration, to be borne in mind, is, that, in this Country, there is much less real need of charity, in supplying physical necessities, than is generally supposed, by those who have not learned the more excellent way. This Land is so abundant in supplies, and labor is in such demand, that every healthy person can earn a comfortable support. And if all the poor were instantly made virtuous, it is probable that there would be no physical wants, which could not readily be supplied by the immediate friends of each sufferer. The sick, the aged, and the orphan, would be the only objects [Pg 178]of charity. In this view of the case, the primary effort, in relieving the poor, should be, to furnish them the means of earning their own support, and to supply them with those moral influences, which are most effectual in securing virtue and industry.

Another point to be attended to, is, the importance of maintaining a system of *associated* charities. There is no point, in which the economy of charity has more improved, than in the present mode of combining many small contributions, for sustaining enlarged and systematic plans of charity. If all the half-dollars, which are now contributed to aid in organized systems of charity, were returned to the donors, to be applied by the agency and discretion of each, thousands and thousands of the treasures, now employed to promote the moral and intellectual wants of mankind, would become entirely useless. In a democracy, like ours, where few are very rich, and the majority are in comfortable circumstances, this collecting and dispensing of drops and rills, is the mode, by which, in imitation of Nature, the dews and showers are to distil on parched and desert lands. And every person, while earning a pittance to unite with many more, may be cheered with the consciousness of sustaining a grand system of operations, which must have the most decided influence, in raising all mankind to that perfect state of society, which Christianity is designed to secure.

Another consideration, relates to the indiscriminate bestowal of charity. Persons, who have taken pains to inform themselves, and who devote their whole time to dispensing charities, unite in declaring, that this is one of the most fruitful sources of indolence, vice, and poverty. From several of these, the writer has learned, that, by their own personal investigations, they have ascertained, that there are large establishments of idle and wicked persons, in most of our cities, who associate together, to support themselves by every species of imposition. They hire large houses, and live in constant rioting, on the means thus obtained. Among them, are[Pg 179] women who have, or who hire the use of, infant children; others, who are blind, or maimed, or deformed, or who can adroitly feign such infirmities, and, by these means of exciting pity, and by artful tales of wo, they collect alms, both in city and country, to spend in all manner of gross and guilty indulgences. Meantime, many persons, finding themselves often duped by impostors, refuse to give at all; and thus many benefactions are withdrawn, which a wise economy in charity would have secured. For this, and other reasons, it is wise and merciful, to adopt the general rule, never to give alms, till we have had some opportunity of knowing how they will be spent. There are exceptions to this, as to every general rule, which a person of discretion can determine. But the practice, so common among benevolent persons, of giving, at least a trifle, to all who ask, lest, perchance, they may turn away some, who are really sufferers, is one, which causes more sin and misery than it cures.

The writer has never known any system for dispensing charity, so successful, as the one which, in many places, has been adopted in connection with the distribution of tracts. By this method, a town or city is divided into districts; and each district is committed to the care of two ladies, whose duty it is, to call on each family and leave a tract, and make that the occasion for entering into conversation, and learning the situation of all residents in the district. By this method, the ignorant, the vicious, and the poor, are discovered, and their physical, intellectual, and moral, wants, are investigated. In some places, where the writer has resided or visited, each person retained the same district, year after year, so that every poor family in the place was under the watch and care of some intelligent and benevolent lady, who used all her influence to secure a proper education for the children, to furnish them with suitable reading, to encourage habits of industry and economy, and to secure regular attendance on public religious instruction. Thus, the rich and the[Pg 180] poor were brought in contact, in a way advantageous to both parties; and, if such a system could be universally adopted, more would be done for the prevention of poverty and vice, than all the wealth of the Nation could avail for their relief. But this plan cannot be successfully carried out, in this manner, unless there is a large proportion of intelligent, benevolent, and self-denying, persons; and the mere distribution of tracts, without the other parts of the plan, is of very little avail.

But there is one species of charity, which needs especial consideration. It is that, which induces us to refrain from judging of the means and the relative charities of other persons. There have been such indistinct notions, and so many different standards of duty, on this subject, that it is rare for two persons to think exactly alike, in regard to the rule of duty. Each person is bound to inquire and judge for himself, as to his own duty or deficiencies; but as both the resources, and the amount of the actual charities, of other men are beyond our ken, it is as indecorous, as it is uncharitable, to sit in judgement on their decisions.

CHAPTER XVI.
ON ECONOMY OF TIME AND EXPENSES.

On Economy of Time.

The value of time, and our obligation to spend every hour for some useful end, are what few minds properly realize. And those, who have the highest sense of their obligations in this respect, sometimes greatly misjudge in their estimate of what are useful and proper modes of employing time. This arises from limited views of the importance of some pursuits, which they would deem frivolous and useless, but which are,[Pg 181] in reality, necessary to preserve the health of body and mind, and those social affections, which it is very important to cherish. Christianity teaches, that, for all the time afforded us, we must give account to God; and that we have no right to waste a single hour. But time, which is spent in rest or amusement, is often as usefully employed, as if it were devoted to labor or devotion. In employing our time, we are to make suitable allowance for sleep, for preparing and taking food, for securing the means of a livelihood, for intellectual improvement, for exercise and amusement, for social enjoyments, and for benevolent and religious duties. And it is the *right apportionment* of time, to these various duties, which constitutes its true economy.

In making this apportionment, we are bound by the same rules, as relate to the use of property. We are to employ whatever portion is necessary to sustain life and health, as the first duty; and the remainder we are so to apportion, that our highest interests, shall receive the greatest allotment, and our physical gratifications, the least.

The laws of the Supreme Ruler, when He became the civil as well as the religious Head of the Jewish theocracy, furnish an example, which it would be well for all attentively to consider, when forming plans for the apportionment of time and property. To properly estimate this example, it must be borne in mind, that the main object of God, was, to preserve His religion among the Jewish nation; and that they were not required to take any means to propagate it among other nations, as Christians are now required to extend Christianity. So low were they, in the scale of civilization and mental developement, that a system, which confined them to one spot, as an agricultural people, and prevented their growing very rich, or having extensive commerce with other nations, was indispensable to prevent their relapsing into the low idolatries and vices of the nations around them.

The proportion of time and property, which every[Pg 182] Jew was required to devote to intellectual, benevolent, and religious purposes, was as follows:

In regard to property, they were required to give one tenth of all their yearly income, to support the Levites, the priests, and the religious service. Next, they were required to give the first fruits of all their corn, wine, oil, and fruits, and the first-born of all their cattle, for the Lord's treasury, to be employed for the priests, the widow, the fatherless, and the stranger. The first-born, also, of their children, were the Lord's, and were to be redeemed by a specified sum, paid into the sacred treasury. Besides this, they were required to bring a freewill offering to God, every time they went up to the three great yearly festivals. In addition to this, regular yearly sacrifices, of cattle and fowls, were required of each family, and occasional sacrifices for certain sins or ceremonial impurities. In reaping their fields, they were required to leave unreaped, for the poor, the corners; not to glean their fields, olive-yards, or vineyards; and, if a sheaf was left, by mistake, they were not to return for it, but leave it for the poor. When a man sent away a servant, he was thus charged: "Furnish him liberally out of thy flock, and out of thy floor, and out of thy wine-press." When a poor man came to borrow money, they were forbidden to deny him, or to take any interest; and if, at the sabbatical, or seventh, year, he could not pay, the debt was to be cancelled. And to this command, is added the significant caution, "Beware that there be not a thought in thy wicked heart, saying, the seventh year, the year of release, is at hand; and thine eye be evil against thy poor brother, and thou givest him nought; and he cry unto the Lord against thee, and it be sin unto thee. Thou shalt surely give him," "because that for this thing the Lord thy God shall bless thee in all thy works, and in all that thou puttest thine hand unto." Besides this, the Levites were distributed through the land, with the intention that they should be instructors and priests in every part of the nation. Thus, one twelfth of the[Pg 183] people were set apart, having no landed property, to be priests and teachers; and the other tribes were required to support them liberally.

In regard to the time taken from secular pursuits, for the support of religion, an equally liberal amount was demanded. In the first place, one seventh part of their time was taken for the weekly sabbath, when no kind of work was to be done. Then the whole nation were required to meet, at the appointed place, three times a year, which, including their journeys, and stay there, occupied eight weeks, or another seventh part of their time. Then the sabbatical year, when no agricultural labor was to be done, took another seventh of their time from their regular pursuits, as they were an agricultural people. This was the amount of time and property demanded by God, simply to sustain religion and morality within the bounds of that nation. Christianity demands the spread of its blessings to all mankind, and so the restrictions laid on the Jews are withheld, and all our wealth and time, not needful for our own best interest, is to be employed in improving the condition of our fellowmen.

In deciding respecting the rectitude of our pursuits, we are bound to aim at some practical good, as the ultimate object. With every duty of this life, our benevolent Creator has connected some species of enjoyment, to draw us to perform it. Thus, the palate is gratified, by performing the duty of nourishing our bodies; the principle of curiosity is gratified, in pursuing useful knowledge; the desire of approbation is gratified, when we perform benevolent and social duties; and every other duty has an alluring enjoyment connected with it. But the great mistake of mankind has consisted in seeking the pleasures, connected with these duties, as the sole aim, without reference to the main end that should be held in view, and to which the enjoyment should be made subservient. Thus, men seek to gratify the palate, without reference to the question whether the body is properly nourished;[Pg 184] and follow after knowledge, without inquiring whether it ministers to good or evil.

But, in gratifying the implanted desires of our nature, we are bound so to restrain ourselves, by reason and conscience, as always to seek the main objects of existence — the highest good of ourselves and others; and never to sacrifice this, for the mere gratification of our sensual desires. We are to gratify appetite, just so far as is consistent with health and usefulness; and the desire for knowledge, just so far as will enable us to do most good by our influence and efforts; and no farther. We are to seek social intercourse, to that extent, which will best promote domestic enjoyment and kindly feelings among neighbors and friends; and we are to pursue exercise and amusement, only so far as will best sustain the vigor of body and mind. For the right apportionment of time, to these and various other duties, we are to give an account to our Creator and final Judge.

Instead of attempting to give any very specific rules on this subject, some modes of economizing time will be suggested. The most powerful of all agencies, in this matter, is, that habit of system and order, in all our pursuits, which has been already pointed out. It is probable, that a regular and systematic employment of time, will enable a person to accomplish thrice the amount of labor, that could otherwise be performed.

Another mode of economizing time, is, by uniting several objects in one employment. Thus, exercise, or charitable efforts, can be united with social enjoyments, as is done in associations for sewing, or visiting the poor. Instruction and amusement can also be combined. Pursuits like music, gardening, drawing, botany, and the like, unite intellectual improvement with amusement, social enjoyment, and exercise.

With housekeepers, and others whose employments are various and desultory, much time can be saved by preparing employments for little intervals of leisure. Thus, some ladies make ready, and keep in the parlor,[Pg 185] light work, to take up when detained there; some keep a book at hand, in the nursery, to read while holding or sitting by a sleeping infant. One of the most popular female poets of our Country very often shows her friends, at their calls, that the thread of the knitting, never need interfere with the thread of agreeable discourse.

It would be astonishing, to one who had never tried the experiment, how much can be accomplished, by a little planning and forethought, in thus finding employment for odd intervals of time.

But, besides economizing our own time, we are bound to use our influence and example to promote the discharge of the same duty by others. A woman is under obligations so to arrange the hours and pursuits of her family, as to promote systematic and habitual industry; and if, by late breakfasts, irregular hours for meals, and other hinderances of this kind, she interferes with, or refrains from promoting regular industry in, others, she is accountable to God for all the waste of time consequent on her negligence. The mere example of system and industry, in a housekeeper, has a wonderful influence in promoting the same virtuous habit in others.

On Economy in Expenses.

It is impossible for a woman to practise a wise economy in expenditures, unless she is taught how to do it, either by a course of experiments, or by the instruction of those who have had experience. It is amusing to notice the various, and oftentimes contradictory, notions of economy, among judicious and experienced housekeepers; for there is probably no economist, who would not be deemed lavish or wasteful, in some respects, by another and equally experienced and judicious person, who, in some different points, would herself be as much condemned by the other. These diversities are occasioned by dissimilar early habits, and by the different relative value assigned, by[Pg 186] each, to the various modes of enjoyment, for which money is expended.

But, though there may be much disagreement in minor matters, there are certain general principles, which all unite in sanctioning. The first, is, that care be taken to know the amount of income and of current expenses, so that the proper relative proportion be preserved, and the expenditures never exceed the means. Few women can do this, thoroughly, without keeping regular accounts. The habits of this Nation, especially among business-men, are so desultory, and the current expenses of a family, in many points, are so much more under the control of the man than of the woman, that many women, who are disposed to be systematic in this matter, cannot follow their wishes. But there are often cases, when much is left undone in this particular, simply because no effort is made. Yet every woman is bound to do as much as is in her power, to accomplish a systematic mode of expenditure, and the regulation of it by Christian principles.

The following are examples of different methods which have been adopted, for securing a proper adjustment of expenses to the means.

The first, is that of a lady, who kept a large boarding-house, in one of our cities. Every evening, before retiring, she took an account of the expenses of the day; and this usually occupied her not more than fifteen minutes, at a time. On each Saturday, she made an inventory of the stores on hand, and of the daily expenses, and also of what was due to her; and then made an exact estimate of her expenditures and profits. This, after the first two or three weeks, never took more than an hour, at the close of the week. Thus, by a very little time, regularly devoted to this object, she knew, accurately, her income, expenditures, and profits.

Another friend of the writer, lives on a regular salary. The method adopted, in this case, is to calculate to what the salary amounts, each week. Then an[Pg 187] account is kept, of what is paid out, each week, for rent, fuel, wages, and food. This amount of each week is deducted from the weekly income. The remainders of each week are added, at the close of a month, as the stock from which is to be taken, the dress, furniture, books, travelling expenses, charities, and all other expenditures.

Another lady, whose husband is a lawyer, divides the year into four quarters, and the income into four equal parts. She then makes her plans, so that the expenses of one quarter shall never infringe on the income of another. So resolute is she, in carrying out this determination, that if, by any mischance, she is in want of articles before the close of a quarter, which she has not the means for providing, she will subject herself to temporary inconvenience, by waiting, rather than violate her rule.

Another lady, whose husband is engaged in a business, which he thinks makes it impossible for him to know what his yearly income

194

will be, took this method:—She kept an account of all her disbursements, for one year. This she submitted to her husband, and obtained his consent, that the same sum should be under her control, the coming year, for similar purposes, with the understanding, that she might modify future apportionments, in any way her judgement and conscience might approve.

A great deal of uneasiness and discomfort is caused, to both husband and wife, in many cases, by an entire want of system and forethought, in arranging expenses. Both keep buying what they think they need, without any calculation as to how matters are coming out, and with a sort of dread of running in debt, all the time harassing them. Such never know the comfort of independence. But, if a man or woman will only calculate what their income is, and then plan so as to know that they are all the time living within it, they secure one of the greatest comforts, which wealth ever bestows, and what many of the rich, who live in a[Pg 188] loose and careless way, never enjoy. It is not so much the amount of income, as the regular and correct apportionment of expenses, that makes a family truly comfortable. A man, with ten thousand a year, is often more harassed, for want of money, than the systematic economist, who supports a family on only six hundred a year. And the inspired command, "Owe no man any thing," can never be conscientiously observed, without a systematic adaptation of expenses to means.

As it is very important that young ladies should learn systematic economy, in expenses, it will be a great benefit, for every young girl to begin, at twelve or thirteen years of age, to make her own purchases, and keep her accounts, under the guidance of her mother, or some other friend. And if parents would ascertain the actual expense of a daughter's clothing, for a year, and give the sum to her, in quarterly payments, requiring a regular account, it would be of great benefit in preparing her for future duties. How else are young ladies to learn to make purchases properly, and to be systematic and economical? The art of system and economy can no more come by intuition, than the art of watchmaking or bookkeeping; and how strange it appears, that so many young ladies take charge of a husband's establishment, without having had either instruction or experience in one of the most important duties of their station!

The second general principle of economy, is, that, in apportioning an income, among various objects, the most important should receive the largest supply, and that all retrenchments be made in matters of less importance. In a previous chapter, some general principles have been presented, to guide in this duty. Some additional hints will here be added, on the same topic.

In regard to dress and furniture, much want of judgement and good taste is often seen, in purchasing some expensive article, which is not at all in keeping with the other articles connected with it. Thus, a[Pg 189] large sideboard, or elegant mirror, or sofa, which would be suitable only for a large establishment, with other rich furniture, is crowded into too small a room, with coarse and cheap articles around it. So, also, sometimes a parlor, and company-chamber, will be furnished in a style suitable only for the wealthy, while the table will be supplied with shabby linen, and imperfect crockery, and every other part of the house will look, in comparison with these fine rooms, mean and niggardly. It is not at all uncommon, to find very showy and expensive articles in the part of the house visible to strangers, when the children's rooms, kitchen, and other back portions, are on an entirely different scale.

So in regard to dress, a lady will sometimes purchase an elegant and expensive article, which, instead of attracting admiration from the eye of taste, will merely serve as a decoy to the painful contrast of all other parts of the dress. A woman of real good taste and discretion, will strive to maintain a relative consistency between all departments, and not, in one quarter, live on a scale fitted only to the rich, and in another, on one appropriate only to the poor.

Another mistake in economy, is often made, by some of the best-educated and most intelligent of mothers. Such will often be found spending day after day at needlework, when, with a comparatively small sum, this labor could be obtained of those who need the money, which such work would procure for them. Meantime, the daughters of the family, whom the mother is qualified to educate, or so nearly qualified, that she could readily keep ahead of her children, are sent to expensive boarding-schools, where their delicate frames, their pliant minds, and their moral and religious interests, are relinquished to the hands of strangers. And the expense, thus incurred, would serve to pay the hire of every thing the mother can do in sewing, four or five times over. The same want of economy is shown in communities, where, instead of establishing a good female school in their vicinity, the[Pg 190] men of wealth send their daughters abroad, at double the expense, to be either educated or spoiled, as the case may be.

Another species of poor economy, is manifested in neglecting to acquire and apply mechanical skill, which, in consequence, has to be hired from others. Thus, all the plain sewing will be done by the mother and daughters, while all that requires skill will be hired. Instead of this, others take pains to have their daughters instructed in mantuamaking, and the simpler parts of millinery, so that the plain work is given to the poor, who need it, and the more expensive and tasteful operations are performed in the family. The writer knows ladies, who not only make their own dresses, but also their caps, bonnets, and artificial flowers.

Some persons make miscalculations in economy, by habitually looking up cheap articles, while others go to the opposite extreme, and always buy the best of every thing. Those ladies, who are considered the best economists, do not adopt either method. In regard to cheap goods, the fading colors, the damages discovered in use, the poorness of material, and the extra sewing demanded to replace articles lost by such causes, usually render them very dear, in the end. On the other hand, though some articles, of the most expensive kind, wear longest and best, yet, as a general rule, articles at medium prices do the best service. This is true of table and bed linens, broadcloths, shirtings, and the like; though, even in these cases, it is often found, that the coarsest and cheapest last the longest.

Buying by wholesale, and keeping a large supply on hand, are economical only in large families, where the mistress is careful; but in other cases, the hazards of accident, and the temptation to a lavish use, will make the loss outrun the profits.

There is one mode of economizing, which, it is hoped, will every year grow more rare; and that is, making penurious savings, by getting the poor to work as cheap[Pg 191] as possible. Many amiable and benevolent women have done this, on principle, without reflecting on the want of Christian charity thus displayed. Let every woman, in making bargains with the poor, conceive herself placed in the same circumstances, toiling hour after hour, and day after day, for a small sum, and then deal with others as she would be dealt by in such a situation. *Liberal prices*, and *prompt payment*, should be an invariable maxim, in dealing with the poor.

The third general principle of economy, is, that all articles should be so used, and taken care of, as to secure the longest service, with the least waste. Under this head, come many particulars in regard to the use and preservation of articles, which will be found more in detail in succeeding chapters. It may be proper, however, here to refer to one very common impression, as to the relative obligation of the poor and the rich in regard to economy. Many seem to suppose, that those who are wealthy, have a right to be lavish and negligent in the care of expenses. But this surely is a great mistake. Property is a talent, given by God, to spend for the welfare of mankind; and the needless waste of it, is as wrong in the rich, as it is in the poor. The rich are under obligations to apportion their income, to the various objects demanding attention, by the same rule as all others; and if this will allow them to spend more for superfluities than those of smaller means, it never makes it right to misuse or waste any of the bounties of Providence. Whatever is no longer wanted for their own enjoyment, should be carefully saved, to add to the enjoyment of others.

It is not always that men understand the economy of Providence, in that unequal distribution of property, which, even under the most perfect form of government, will always exist. Many, looking at the present state of things, imagine that the rich, if they acted in strict conformity to the law of benevolence, would share all their property with their suffering fellow-men. But such do not take into account, the inspired declaration,[Pg 192] that "a man's life consisteth not in the abundance of the things which he possesseth," or, in other words, life is made valuable, not by great possessions, but by such a character as prepares a man to enjoy what he holds. God perceives that human character can be most improved, by that kind of discipline, which exists, when there is something valuable to be gained by industrious efforts. This stimulus to industry could never exist, in a community where all are just alike, as it does in a state of society where every man sees, possessed by others, enjoyments, which he desires, and may secure by effort and industry. So, in a community where all are alike as to property, there would be no chance to gain that noblest of all attainments, a habit of self-denying benevolence, which toils for the good of others, and takes from one's own store, to increase the enjoyments of another.

Instead, then, of the stagnation, both of industry and of benevolence, which would follow the universal and equable distribution of property, one class of men, by superior advantages of birth, or intellect, or patronage, come into possession of a great amount of capital. With these means, they are enabled, by study, reading, and travel, to secure expansion of mind, and just views of the relative advantages of moral, intellectual, and physical enjoyments. At the same time, Christianity imposes obligations, corresponding with the increase of advantages and means. The rich are not at liberty to spend their treasures for themselves, alone. Their wealth is given, by God, to be employed for the best good of mankind; and their intellectual advantages are designed, primarily, to enable them to judge correctly, in employing their means most wisely for the general good.

Now, suppose a man of wealth inherits ten thousand acres of real estate: it is not his duty to divide it among his poor neighbors and tenants. If he took this course, it is probable, that most of them would spend all in thriftless waste and indolence, or in mere[Pg 193] physical enjoyments. Instead, then, of thus putting his capital out of his hands, he is bound to retain, and so to employ, it, as to raise his neighbors and tenants to such a state of virtue and intelligence, that they can secure far more, by their own efforts and industry, than he, by dividing his capital, could bestow upon them.

In this view of the subject, it is manifest, that the unequal distribution of property is no evil. The great difficulty is, that so large a portion of those who hold much capital, instead of using their various advantages for the greatest good of those around them, employ the chief of them for mere selfish indulgences; thus inflicting as much mischief on themselves, as results to others from their culpable neglect. A great portion of the rich seem to be acting on the principle, that the more God bestows on them, the less are they under obligation to practise any self-denial, in fulfilling his benevolent plan of raising our race to intelligence and holiness.

There are not a few, who seem to imagine that it is a mark of gentility to be careless of expenses. But this notion, is owing to a want of knowledge of the world. As a general fact, it will be found, that persons of rank and wealth, abroad, are much more likely to be systematic and economical, than persons of inferior standing in these respects. Even the most frivolous, among the rich and great, are often found practising a rigid economy, in certain respects, in order to secure gratifications in another direction. And it will be found so common, among persons of vulgar minds, and little education, and less sense, to make a display of profusion and indifference to expense, as a mark of their claims to gentility, that the really genteel look upon it rather as a mark of low breeding. So that the sort of feeling, which some persons cherish, as if it were a degradation to be careful of small sums, and to be attentive to relative prices, in making purchases, is founded on mistaken notions of gentility and propriety.

[Pg 194]

But one caution is needful, in regard to another extreme. When a lady of wealth, is seen roaming about in search of cheaper articles, or trying to beat down a shopkeeper, or making a close bargain with those she employs, the impropriety is glaring to all minds. A person of wealth has no occasion to spend time in looking for extra cheap articles; her time could be more profitably employed in distributing to the wants of others. And the practice of beating down tradespeople, is vulgar and degrading, in any one. A woman, after a little inquiry, can ascertain what is the fair and common price of things; and if she is charged an exorbitant sum, she can decline taking the article. If the price be a fair one, it is not becoming in her to search for another article which is below the regular charge. If a woman finds that she is in a store where they charge high prices, expecting to be beat down, she can mention, that she wishes to know the lowest price, as it is contrary to her principles to beat down charges.

There is one inconsistency, worthy of notice, which is found among that class, who are ambitious of being ranked among the aristocracy of society. It has been remarked, that, in the real aristocracy of other lands, it is much more common, than with us, to practise systematic economy. And such do not hesitate to say so, when they cannot afford certain indulgences. This practice descends to subordinate grades; so that foreign ladies, when they come to reside among us, seldom hesitate in assigning the true reason, when they cannot afford any gratification. But in this Country, it will be found, that many, who are most fond of copying aristocratic examples, are, on this point, rather with the vulgar. Not a few of those young persons, who begin life with parlors and dresses in a style fitting only to established wealth, go into expenses, which they can ill afford; and are ashamed even to allow, that they are restrained from any expense, by motives of economy. Such a confession is never extorted,[Pg 195] except by some call of benevolence; and then, they are very ready to declare that they cannot afford to bestow even a pittance. In such cases, it would seem as if the direct opposite of Christianity had gained possession of their tastes and opinions. They are ashamed to appear to deny themselves; but are very far from having any shame in denying the calls of benevolence.

CHAPTER XVII.
ON HEALTH OF MIND.

There is such an intimate connection between the body and mind, that the health of one, cannot be preserved, without a proper care of the other. And it is from a neglect of this principle, that some of the most exemplary and conscientious persons in the world, suffer a thousand mental agonies, from a diseased state of body, while others ruin the health of the body, by neglecting the proper care of the mind. When the brain is excited, by stimulating drinks taken into the stomach, it produces a corresponding excitement of the mental faculties. The reason, the imagination, and all the powers, are stimulated to preternatural vigor and activity. In like manner, when the mind is excited by earnest intellectual effort, or by strong passions, the brain is equally excited, and the blood rushes to the head. Sir Astley Cooper records, that, in examining the brain of a young man who had lost a portion of his skull, whenever "he was agitated, by some opposition to his wishes," "the blood was sent, with increased force, to his brain," and the pulsations "became frequent and violent." The same effect was produced by any intellectual effort; and the flushed countenance, which attends earnest study or strong emotions of fear, shame, or anger, is an external indication of the suffused state of the brain from such causes.

[Pg 196]

In exhibiting the causes, which injure the health of the mind, they will be found to be partly physical, partly intellectual, and partly moral.

The first cause of mental disease and suffering, is not unfrequently found in the want of a proper supply of duly oxygenized blood. It has been shown, that the blood, in passing through the lungs, is purified, by the oxygen of the air combining with the superabundant hydrogen and carbon of the venous blood, thus forming carbonic acid and water, which are expired into the atmosphere. Every pair of lungs is constantly withdrawing from the surrounding atmosphere its healthful principle, and returning one, which is injurious to human life.

When, by confinement, and this process, the atmosphere is deprived of its appropriate supply of oxygen, the purification of the blood is interrupted, and it passes, without being properly prepared, into the brain, producing languor, restlessness, and inability to exercise the intellect and feelings. Whenever, therefore, persons sleep in a close apartment, or remain, for a length of time, in a crowded or ill-ventilated room, a most pernicious influence is exerted on the brain, and, through this, on the mind. A person, who is often exposed to such influences, can never enjoy that elasticity and vigor of mind, which is one of the chief indications of its health. This is the reason, why all rooms for religious meetings, and all schoolrooms, and sleeping apartments, should be so contrived, as to secure a constant supply of fresh air from without. The minister, who preaches in a crowded and ill-ventilated apartment, loses much of his power to feel and to speak, while the audience are equally reduced, in their capability of attending. The teacher, who confines children in a close apartment, diminishes their ability to study, or to attend to his instructions. And the person, who habitually sleeps in a close room, impairs his mental energies, in a similar degree. It is not unfrequently the case, that depression of spirits, and stupor of intellect, are occasioned solely by inattention to this subject.

[Pg 197]

Another cause of mental disease, is, the excessive exercise of the intellect or feelings. If the eye is taxed, beyond its strength, by protracted use, its blood-vessels become gorged, and the bloodshot appearance warns of the excess and the need of rest. The brain is affected, in a similar manner, by excessive use, though the suffering and inflamed organ cannot make its appeal to the eye. But there are some indications, which ought never to be misunderstood or disregarded. In cases of pupils, at school or at college, a diseased state, from over action, is often manifested by increased clearness of mind, and ease and vigor of mental action. In one instance, known to the writer, a most exemplary and industrious pupil, anxious to improve every hour, and ignorant or unmindful of the laws of health, first manifested the diseased state of her brain and mind, by demands for more studies, and a sudden and earnest activity in planning modes of improvement for herself and others. When warned of her danger, she protested that she never was better, in her life; that she took regular exercise, in the open air, went to bed in season, slept soundly, and felt perfectly well; that her mind was never before so bright and clear, and study never so easy and delightful. And at this time, she was on the verge of derangement, from which she was saved only by an entire cessation of all her intellectual efforts.

A similar case occurred, under the eye of the writer, from over-excited feelings. It was during a time of unusual religious interest in the community, and the mental disease was first manifested, by the pupil bringing her Hymn-book or Bible to the class-room, and making it her constant resort, in every interval of school duty. It finally became impossible to convince her, that it was her duty to attend to any thing else; her conscience became morbidly sensitive, her perceptions indistinct, her deductions unreasonable, and nothing, but entire change of scene, exercise, and amusement, saved her. When the health of the brain[Pg 198] was restored, she found that she could attend to the "one thing needful," not only without interruption of duty, or injury of health, but rather so as to promote both. Clergymen and teachers need most carefully to notice and guard against the danger here alluded to.

Any such attention to religion, as prevents the performance of daily duties and needful relaxation, is dangerous, as tending to produce such a state of the brain, as makes it impossible to feel or judge correctly. And when any morbid and unreasonable pertinacity appears, much exercise, and engagement in other interesting pursuits, should be urged, as the only mode of securing the religious benefits aimed at. And whenever any mind is oppressed with care, anxiety, or sorrow, the amount of active exercise in the fresh air should be greatly increased, that the action of the muscles may withdraw the blood, which, in such seasons, is constantly tending too much to the brain.

There has been a most appalling amount of suffering, derangement, disease, and death, occasioned by a want of attention to this subject, in teachers and parents. Uncommon precocity in children is usually the result of an unhealthy state of the brain; and, in such cases, medical men would now direct, that the wonderful child should be deprived of all books and study, and turned to play or work in the fresh air. Instead of this, parents frequently add fuel to the fever of the brain, by supplying constant mental stimulus, until the victim finds refuge in idiocy or an early grave. Where such fatal results do not occur, the brain, in many cases, is so weakened, that the prodigy of infancy sinks below the medium of intellectual powers in afterlife. In our colleges, too, many of the most promising minds sink to an early grave, or drag out a miserable existence, from this same cause. And it is an evil, as yet little alleviated by the increase of physiological knowledge. Every college and professional school, and every seminary for young ladies, needs a medical man, not only to lecture on physiology and the laws of health, but empowered, in his official[Pg 199] capacity, to investigate the case of every pupil, and, by authority, to restrain him to such a course of study, exercise, and repose, as his physical system requires. The writer has found, by experience, that, in a large institution, there is one class of pupils who need to be restrained, by penalties, from late hours and excessive study, as much as another class need stimulus to industry.

Under the head of excessive mental action, must be placed the indulgence of the imagination in *novel reading* and *castle building*. This kind of stimulus, unless counterbalanced by physical exercise, not only wastes time and energies, but undermines the vigor of the nervous system. The imagination was designed, by our kind Creator, as the charm and stimulus to animate to benevolent activity; and its perverted exercise seldom fails to bring the appropriate penalty.

A third cause of mental disease, is, the want of the appropriate exercise of the various faculties of the mind. On this point, Dr. Combe remarks, "We have seen, that, by disuse, muscle becomes emaciated, bone softens, blood-vessels are obliterated, and nerves lose their characteristic structure. The brain is no exception to this general rule. Of it, also, the tone is impaired by permanent inactivity, and it becomes less fit to manifest the mental powers with readiness and energy." It is "the withdrawal of the stimulus necessary for its healthy exercise, which renders solitary confinement so severe a punishment, even to the most daring minds. It is a lower degree of the same cause, which renders continuous seclusion from society so injurious, to both mental and bodily health."

"*Inactivity of intellect and of feeling* is a very frequent predisposing cause of every form of nervous disease. For demonstrative evidence of this position, we have only to look at the numerous victims to be found, among persons who have no call to exertion in gaining the means of subsistence, and no objects of interest on which to exercise their mental faculties[Pg 200] and who consequently sink into a state of mental sloth and nervous weakness." "If we look abroad upon society, we shall find innumerable examples of mental and nervous debility from this cause. When a person of some mental capacity is confined, for a long time, to an unvarying round of employment, which affords neither scope nor stimulus for one half of his faculties, and, from want of education or society, has no external resources; his mental powers, for want of exercise, become blunted, and his perceptions slow and dull." "The intellect and feelings, not being provided with interests external to themselves, must either become inactive and weak, or work upon themselves and become diseased."

"The most frequent victims of this kind of predisposition, are females of the middle and higher ranks, especially those of a nervous constitution and *good natural abilities*; but who, from an ill-directed education, possess nothing more solid than mere accomplishments, and have no materials of thought," and no "occupation to excite interest or *demand* attention." "The liability of such persons to melancholy, hysteria, hypochondriasis, and other varieties of mental distress, really depends on a state of irritability of brain, induced by imperfect exercise."

These remarks, of a medical man, illustrate the principles before indicated;—namely, that the demand of Christianity, that we live to promote the general happiness, and not merely for selfish indulgence, has for its aim, not only the general good, but the highest happiness, of the individual of whom it is required.

A person possessed of wealth, who has nothing more noble to engage his attention, than seeking his own personal enjoyment, subjects his mental powers and moral feelings to a degree of inactivity, utterly at war with health of mind. And the greater the capacities, the greater are the sufferings which result from this cause. Any one, who has read the misanthropic wailings of Lord Byron, has seen the necessary result of[Pg 201] great and noble powers bereft of their appropriate exercise, and, in consequence, becoming sources of the keenest suffering.

It is this view of the subject, which has often awakened feelings of sorrow and anxiety in the mind of the writer, while aiding in the developement and education of superior female minds, in the wealthier circles. Not because there are not noble objects for interest and effort, abundant, and within reach of such minds; but because long-established custom has made it seem so Quixotic, to the majority, even of the professed followers of Christ, for a woman of wealth to practise any great self-denial, that few have independence of mind and Christian principle sufficient to overcome such an influence. The more a mind has its powers developed, the more does it aspire and pine after some object worthy of its energies and affections; and they are commonplace and phlegmatic characters, who are most free from such deep-seated wants. Many a young woman, of fine genius and elevated sentiment, finds a charm in Lord Byron's writings, because they present a glowing picture of what, to a certain extent, must be felt by every well-developed mind, which has no nobler object in life, than the pursuit of its own gratification.

If young ladies of wealth could pursue their education, under the full conviction that the increase of their powers and advantages increased their obligations to use all for the good of society, and with some plan of benevolent enterprise in view, what new motives of interest would be added to their daily pursuits! And what blessed results would follow, to our beloved Country, if all well-educated females carried out the principles of Christianity, in the exercise of their developed powers!

It is cheering to know, that there are women, among the most intelligent and wealthy, who can be presented as examples of what may be done, when there is a heart to do. A pupil of the writer is among this number, [Pg 202]who, though a rich heiress, immediately, on the close of her school-life, commenced a course of self-denying benevolence, in the cause of education. She determined to secure a superior female institution, in her native place, which should extend the benefits of the best education to all in that vicinity, at a moderate charge. Finding no teacher on the ground, prepared to take the lead, and though herself a timid and retiring character, she began, with the aid of the governess in her mother's family, a daily school, superintending all, and teaching six hours a day. The liberal-minded and intelligent mother cooperated, and the result is a flourishing female seminary, with a large and beautiful and well-furnished building; the greater part of the means being supplied by the mother, and almost all by the members of that family connection. And both these ladies will testify, that no time or money, spent for any other object, has ever secured to them more real and abiding enjoyment, than witnessing the results of this successful and benevolent enterprise, which, for years to come, will pour forth blessings on society.

Another lady could be pointed out, who, possessing some property, went into a new western village, built and furnished her schoolhouse, and established herself there, to aid in raising a community from ignorance and gross worldliness, to intelligence and virtue. And in repeated instances, among the friends and pupils of the writer, young ladies have left wealthy homes, and affectionate friends, to find nobler enjoyments, in benevolent and active exertions to extend intelligence and virtue, where such disinterested laborers were needed. In other cases, where it was not practicable to leave home, well-educated young ladies have interested themselves in common schools in the vicinity, aiding the teachers, by their sympathy, counsel, and personal assistance.

Other ladies, of property and standing, having families to educate, and being well qualified for such duties, have relinquished a large portion of domestic labor[Pg 203] and superintendence, which humbler minds could be hired to perform, devoted themselves to the education of their children, and received others, less fortunate, to share with their own these superior advantages. But, so long as the feeling widely exists, that the increase of God's bounties diminishes the obligations of self-denying service for the good of mankind, so long will well-educated women, in easy circumstances, shrink from such confinement and exertion.

It is believed, however, that there are many benevolent and intelligent women, in this Country, who would gladly engage in such enterprises, were there any appropriate way within their reach. And it is a question, well deserving consideration, among those who guide the public mind in benevolent enterprises, whether some organization is not demanded, which shall bring the whole community to act systematically, in voluntary associations, to extend a proper education to every child in this Nation, and to bring into activity all the female enterprise and benevolence now lying dormant, for want of proper facilities to exercise them. There are hundreds of villages, which need teachers, and that would support them, if they were on the spot, but which never will send for them. And there are hundreds of females, now unemployed, who would teach, if a proper place, and home, and support, and escort, were provided for them. And there needs to be some enlarged and systematic plan, conducted by wise and efficient men, to secure these objects.

Could such a plan, as the one suggested, be carried out, it is believed that many female minds, now suffering, from diseases occasioned by want of appropriate objects for their energies, would be relieved. The duties of a teacher exercise every intellectual faculty, to its full extent; while, in this benevolent service, all the social, moral, and benevolent, emotions, are kept in full play. The happiest persons the writer has ever known, — those who could say that they were as happy as they wished to be, in this world, (and she has seen such,) — were persons engaged in this employment.

[Pg 204]

The indications of a diseased mind, owing to a want of the proper exercise of its powers, are, apathy, discontent, a restless longing for excitement, a craving for unattainable good, a diseased and morbid action of the imagination, dissatisfaction with the world, and factitious interest in trifles which the mind feels to be unworthy of its powers. Such minds sometimes seek alleviation in exciting amusements; others resort to the grosser enjoyments of sense. Oppressed with the extremes of languor, or over-excitement, or apathy, the body fails under the wearing process, and adds new causes of suffering to the mind. Such, the compassionate Saviour calls to his service, in these appropriate terms: "Come unto Me, all ye that labor and are heavy laden, and I will give you rest. Take My yoke upon you, and learn of Me," "and ye shall find rest unto your souls."

CHAPTER XVIII.
ON THE CARE OF DOMESTICS.

There is no point, where the women of this Country need more wisdom, patience, principle, and self-control, than in relation to those whom they employ in domestic service. The subject is attended with many difficulties, which powerfully influence the happiness of families; and the following suggestions are offered, to aid in securing right opinions and practice.

One consideration, which it would be well to bear in mind, on this subject, is, that a large portion of the peculiar trials, which American women suffer from this source, are the necessary evils connected with our most valuable civil blessings. Every blessing of this life involves some attendant liability to evil, from the same source; and, in this case, while we rejoice at a state of society, which so much raises the condition and advantages of our sex, the evils involved should be regarded[Pg 205] as more than repaid, by the compensating benefits. If we cannot secure the cringing, submissive, well-trained, servants of aristocratic lands, let us be consoled that we thus escape from the untold miseries and oppression, which always attend that state of society.

Instead, then, of complaining that we cannot have our own peculiar advantages, and those of other nations, too, or imagining how much better off we should be, if things were different from what they are, it is much wiser and more Christianlike to strive cheerfully to conform to actual circumstances; and, after remedying all that we can control, patiently to submit to what is beyond our power. If domestics are found to be incompetent, unstable, and unconformed to their station, it is Perfect Wisdom which appoints these trials, to teach us patience, fortitude, and self-control; and, if the discipline is met, in a proper spirit, it will prove a blessing, rather than an evil.

But, to judge correctly in regard to some of the evils involved in the state of domestic service, in this Country, we should endeavor to conceive ourselves placed in the situation of those, of whom complaint is made, that we may not expect, from them, any more than it would seem right should be exacted from us, in similar circumstances.

It is sometimes urged, against domestics, that they exact exorbitant wages. But what is the rule of rectitude, on this subject? Is it not the universal law of labor and of trade, that an article is to be valued, according to its scarcity and the demand? When wheat is scarce, the farmer raises his price; and when a mechanic offers services, difficult to be obtained, he makes a corresponding increase of price. And why is it not right, for domestics to act according to a rule, allowed to be correct in reference to all other trades and professions? It is a fact, that really good domestic service must continue to increase in value, just in proportion as this Country waxes rich and prosperous; thus making the proportion of those, who wish to hire labor, relatively [Pg 206]greater, and the number of those, willing to go to service, less.

Money enables the rich to gain many advantages, which those of more limited circumstances cannot secure. One of these, is, securing good domestics, by offering high wages; and this, as the scarcity of this class increases, will serve constantly to raise the price of service. It is right for domestics to charge the market value, and this value is always decided by the scarcity of the article and the amount of demand. Right views of this subject, will sometimes serve to diminish hard feelings towards those, who would otherwise be wrongfully regarded as unreasonable and exacting.

Another complaint against domestics, is, that of instability and discontent, leading to perpetual change. But in reference to this, let a mother or daughter conceive of their own circumstances as so changed, that the daughter must go out to service. Suppose a place is engaged, and it is then found that she must sleep in a comfortless garret; and that, when a new domestic comes, perhaps a coarse and dirty foreigner, she must share her bed with her. Another place is offered, where she can have a comfortable room, and an agreeable room-mate; in such a case, would not both mother and daughter think it right to change?

Or, suppose, on trial, it was found that the lady of the house was fretful, or exacting, and hard to please; or, that her children were so ungoverned, as to be perpetual vexations; or, that the work was so heavy, that no time was allowed for relaxation and the care of a wardrobe;—and another place offers, where these evils can be escaped: would not mother and daughter here think it right to change? And is it not right for domestics, as well as their employers, to seek places, where they can be most comfortable?

In some cases, this instability and love of change would be remedied, if employers would take more pains to make a residence with them agreeable; and to attach[Pg 207] domestics to the family, by feelings of gratitude and affection. There are ladies, even where well-qualified domestics are most rare, who seldom find any trouble in keeping good and steady ones. And the reason is, that their domestics know they cannot better their condition, by any change within reach. It is not merely by giving them comfortable rooms, and good food, and presents, and privileges, that the attachment of domestics is secured; it is by the manifestation of a friendly and benevolent interest in their comfort and improvement. This is exhibited, in bearing patiently with their faults; in kindly teaching them how to improve; in showing them how to make and take proper care of their clothes; in guarding their health; in teaching them to read, if necessary, and supplying them with proper books; and, in short, by endeavoring, so far as may be, to supply the place of parents. It is seldom that such a course would fail to secure steady service, and such affection and gratitude, that even higher wages would be ineffectual to tempt them away. There would probably be some cases of ungrateful returns; but there is no doubt that the course indicated, if generally pursued, would very much lessen the evil in question.

Another subject of complaint, in regard to domestics, is, their pride, insubordination, and spirit not conformed to their condition. They are not willing to be called *servants*; in some places, they claim a seat, at meals, with the family; they imitate a style of dress unbecoming their condition; and their manners and address are rude and disrespectful. That these evils are very common, among this class of persons, cannot be denied; the only question is, how can they best be met and remedied.

In regard to the common feeling among domestics, which is pained and offended by being called "servants," there is need of some consideration and allowance. It should be remembered, that, in this Country, children, from their earliest years, are trained to abhor[Pg 208] slavery, in reference to themselves, as the greatest of all possible shame and degradation. They are perpetually hearing orations, songs, and compositions of all sorts, which set forth the honor and dignity of freemen, and heap scorn and contempt on all who would be so mean as to be slaves. Now the term servant, and the duties it involves, are, in the minds of many persons, nearly the same as those of slave. And there are few minds, entirely free from associations which make servitude a degradation. It is not always pride, then, which makes this term so offensive. It is a consequence of that noble and generous spirit of freedom, which every American draws from his mother's breast, and which ought to be respected, rather than despised. In order to be respected, by others, we must respect ourselves; and sometimes the ruder classes of society make claims, deemed forward and offensive, when, with their views, such a position seems indispensable to preserve a proper self-respect.

Where an excessive sensibility on this subject exists, and forward and disrespectful manners result from it, the best remedy is, a kind attempt to give correct views, such as better-educated minds are best able to attain. It should be shown to them, that, in this Country, labor has ceased to be degrading, in any class; that, in all classes, different grades of subordination must exist; and that it is no more degrading, for a domestic to regard the heads of a family as superiors in station, and treat them with becoming respect, than it is for children to do the same, or for men to treat their rulers with respect and deference. They should be taught, that domestics use a different entrance to the house, and sit at a distinct table, not because they are inferior beings, but because this is the best method of securing neatness, order, and convenience. They can be shown, if it is attempted in a proper spirit and manner, that these very regulations really tend to their own ease and comfort, as well as to that of the family.

The writer has known a case, where the lady of the[Pg 209] family, for the sake of convincing her domestic of the truth of these views, allowed her to follow her own notions, for a short time, and join the family at meals. It was merely required, as a condition, that she should always dress her hair as the other ladies did, and appear in a clean dress, and abide by all the rules of propriety at table, which the rest were required to practise, and which were duly detailed. The experiment was tried, two or three times; and, although the domestic was treated with studious politeness and kindness, she soon felt that she should be much more comfortable in the kitchen, where she could talk, eat, and dress, as she pleased. A reasonable domestic can also be made to feel the propriety of allowing opportunity for the family to talk freely of their private affairs, when they meet at meals, as they never could do, if restrained by the constant presence of a stranger. Such views, presented in a kind and considerate manner, will often entirely change the views of a domestic, who is sensitive on such subjects.

When a domestic is forward and bold in manners, and disrespectful in address, a similar course can be pursued. It can be shown, that those, who are among the best-bred and genteel, have courteous and respectful manners and language to all they meet, while many, who have wealth, are regarded as vulgar, because they exhibit rude and disrespectful manners. The very term, *gentle*man, indicates the refinement and delicacy of address, which distinguishes the high-bred from the coarse and vulgar.

In regard to appropriate dress, in most cases it is difficult for an employer to interfere, *directly*, with comments or advice. The most successful mode, is, to offer some service in mending or making a wardrobe, and when a confidence in the kindness of feeling is thus gained, remarks and suggestions will generally be properly received, and new views of propriety and economy can be imparted. In some cases, it may be well for an employer, — who, from appearances, anticipates [Pg 210]difficulty of this kind, — in making the agreement, to state that she wishes to have the room, person, and dress of her domestics kept neat, and in order, and that she expects to remind them of their duty, in this particular, if it is neglected. Domestics are very apt to neglect the care of their own chambers and clothing; and such habits have a most pernicious influence on their wellbeing, and on that of their children in future domestic life. An employer, then, is bound to exercise a parental care over them, in these respects.

In regard to the great deficiencies of domestics, in qualifications for their duties, much patience and benevolence are required. Multitudes have never been taught to do their work properly; and, in such cases, how unreasonable it would be to expect it of them! Most persons, of this class, depend, for their knowledge in domestic affairs, not on their parents, who are usually unqualified to instruct them, but on their employers; and if they live in a family where nothing is done neatly and properly, they have no chance to learn how to perform their duties well. When a lady finds that she must employ a domestic who is ignorant, awkward, and careless, her first effort should be, to make all proper allowance for past want of instruction, and the next, to remedy the evil, by kind and patient teaching. In doing this, it should ever be borne in mind, that nothing is more difficult, than to change old habits, and to learn to be thoughtful and considerate. And a woman must make up her mind to tell the same thing "over and over again," and yet not lose her patience. It will often save much vexation, if, on the arrival of a new domestic, the mistress of the family, or a daughter, will, for two or three days, go round with the novice, and show the exact manner in which it is expected the work will be done. And this, also, it may be well to specify in the agreement, as some domestics would otherwise resent such a supervision.

But it is often remarked, that, after a woman has taken all this pains to instruct a domestic, and make[Pg 211] her a good one, some other person will offer higher wages, and she will leave. This, doubtless, is a sore trial; but, if such efforts were made in the true spirit of benevolence, the lady will still have her reward, in the consciousness that she has contributed to the welfare of society, by making one more good domestic, and one more comfortable family where that domestic is employed; and if the latter becomes the mother of a family, a whole circle of children will share in the benefit.

There is one great mistake, not unfrequently made, in the management both of domestics and of children; and that is, in supposing that the way to cure defects, is by finding fault as each failing occurs. But, instead of this being true, in many cases the directly opposite course is the best; while, in all instances, much good judgement is required, in order to decide when to notice faults, and when to let them pass unnoticed. There are some minds, very sensitive, easily discouraged, and infirm of purpose. Such persons, when they have formed habits of negligence, haste, and awkwardness, often need expressions of sympathy and encouragement, rather than reproof. They have usually been found fault with, so much, that they have become either hardened or desponding; and it is often the case, that a few words of commendation will awaken fresh efforts and renewed hope. In almost every case, words of kindness, confidence, and encouragement, should be mingled with the needful admonitions or reproof.

It is a good rule, in reference to this point, to *forewarn*, instead of finding fault. Thus, when a thing has been done wrong, let it pass unnoticed, till it is to be done again; and then, a simple request, to have it done in the right way, will secure quite as much, and probably more, willing effort, than a reproof administered for neglect. Some persons seem to take it for granted, that young and inexperienced minds are bound to have all the forethought and discretion of mature persons; and freely express wonder and disgust, when mishaps[Pg 212] occur for want of these traits. But it would be far better to save from mistake or forgetfulness, by previous caution and care on the part of those who have gained experience and forethought; and thus many occasions of complaint and ill-humor will be avoided.

Those, who fill the places of heads of families, are not very apt to think how painful it is, to be chided for neglect of duty, or for faults of character. If they would sometimes imagine themselves in the place of those whom they control, with some person daily administering reproof to them, in the same *tone and style* as they employ to those who are under them, it might serve as a useful check to their chidings. It is often the case, that persons, who are most strict and exacting, and least able to make allowances and receive palliations, are themselves peculiarly sensitive to any thing which implies that they are in fault. By such, the spirit implied in the Divine petition, "forgive us our trespasses as we forgive those who trespass against us," needs especially to be cherished.

One other consideration, is very important. There is no duty, more binding on Christians, than that of patience and meekness under provocations and disappointment. Now, the tendency of every sensitive mind, when thwarted in its wishes, is, to complain and find fault, and that often in tones of fretfulness or anger. But there are few domestics, who have not heard enough of the Bible, to know that angry or fretful fault-finding, from the mistress of a family, when her work is not done to suit her, is not in agreement with the precepts of Christ. They notice and feel the inconsistency; and every woman, when she gives way to feelings of anger and impatience, at the faults of those around her, lowers herself in their respect, while her own conscience, unless very much blinded, cannot but suffer a wound.

There are some women, who, in the main, are amiable, who seem impressed with the idea, that it is their office and duty to find fault with their domestics,[Pg 213] whenever any thing is not exactly right, and follow their fancied calling without the least appearance of tenderness or sympathy, as if the objects of their discipline were stocks or stones. The writer once heard a domestic, describing her situation in a family which she had left, make this remark of her past employer: "She was a very good housekeeper, allowed good wages, and gave us many privileges and presents; but if we ever did any thing wrong, she always *talked to us just as if she thought we had no feelings*, and I never was so unhappy in my life, as while living with her." And this was said of a kind-hearted and conscientious woman, by a very reasonable and amiable domestic.

Every woman, who has the care of domestics, should cultivate a habit of regarding them with that sympathy and forbearance, which she would wish for herself or her daughters, if deprived of parents, fortune, and home. The fewer advantages they have enjoyed, and the greater difficulties of temper or of habit they have to contend with, the more claims they have on compassionate forbearance. They ought ever to be looked upon, not as the mere ministers to our comfort and convenience, but as the humbler and more neglected children of our Heavenly Father, whom He has sent to claim our sympathy and aid.[N]

FOOTNOTE:

[N] The excellent little work of Miss Sedgwick, entitled 'Live, and Let Live,' contains many valuable and useful hints, conveyed in a most pleasing narrative form, which every housekeeper would do well to read. The writer also begs leave to mention a work of her own, entitled, 'Letters to Persons engaged in Domestic Service.'

CHAPTER XIX.
ON THE CARE OF INFANTS.

Every young lady ought to learn how to take proper care of an infant; for, even if she is never to become the responsible guardian of a nursery, she will often [Pg 214]be in situations where she can render benevolent aid to others, in this most fatiguing and anxious duty.

The writer has known instances, in which young ladies, who, having been trained, by their mothers, properly to perform this duty, were, in some cases, the means of saving the lives of infants, and in others, of relieving, by their benevolent aid, sick mothers, from intolerable care and anguish.

On this point, Dr. Combe remarks, "All women are not destined, in the course of Nature, to become mothers; but how very small is the number of those, who are unconnected, by family ties, friendship, or sympathy, with the children of others! How very few are there, who, at some time or other of their lives, would not find their usefulness and happiness increased, by the possession of a kind of knowledge, intimately allied to their best feelings and affections! And how important is it, to the mother herself, that her efforts should be seconded by intelligent, instead of ignorant, assistants!"

In order to be prepared for such benevolent ministries, every young lady should improve the opportunity, whenever it is afforded her, for learning how to wash, dress, and tend, a young infant; and whenever she meets with such a work as Dr. Combe's, on the management of infants, she ought to read it, and *remember* its contents.

It was the design of the author, to fill this chapter chiefly with extracts from various medical writers, giving some of the most important directions on this subject; but finding these extracts too prolix for a work of this kind, she has condensed them into a shorter compass. Some are quoted verbatim, and some are abridged, chiefly from the writings of Doctors Combe, Bell, and Eberle, who are among the most approved writers on this subject.

"Nearly one half of the deaths, occurring during the first two years of existence, are ascribable to mismanagement, and to errors in diet. At birth, the stomach is feeble, and as yet unaccustomed to food; its cravings[Pg 215] are consequently easily satisfied, and frequently renewed." "At that early age, there ought to be no fixed time for giving nourishment. The stomach cannot be thus satisfied." "The active call of the infant, is a sign, which needs never be mistaken."

But care must be taken to determine between the crying of pain or uneasiness, and the call for food; and the practice of giving an infant food, to stop its cries, is often the means of increasing its sufferings. After a child has satisfied its hunger, from two to four hours should intervene, before another supply is given.

"At birth, the stomach and bowels, never having been used, contain a quantity of mucous secretion, which requires to be removed. To effect this, Nature has rendered the first portions of the mother's milk purposely watery and laxative. Nurses, however, distrusting Nature, often hasten to administer some active purgative; and the consequence often is, irritation in the stomach and bowels, not easily subdued." It is only where the child is deprived of its mother's milk, as the first food, that some gentle laxative should be given.

"It is a common mistake, to suppose, that, because a woman is nursing, she ought to live very fully, and to add an allowance of wine, porter, or other fermented liquor, to her usual diet. The only result of this plan, is, to cause an unnatural fulness in the system, which places the nurse on the brink of disease, and retards, rather than increases, the food of the infant. More will be gained by the observance of the ordinary laws of health, than by any foolish deviation, founded on ignorance."

There is no point, on which medical men so emphatically lift the voice of warning, as in reference to administering medicines to infants. It is so difficult to discover what is the matter with an infant, its frame is so delicate and so susceptible, and slight causes have such a powerful influence, that it requires the utmost skill and judgement to ascertain what would be proper medicines, and the proper quantity to be given.

[Pg 216]

Says Dr. Combe, "That there are cases, in which active means must be promptly used, to save the child, is perfectly true. But it is not less certain, that these are cases, of which no mother or nurse ought to attempt the treatment. As a general rule, where the child is well managed, medicine, of any kind, is very rarely required; and if disease were more generally regarded in its true light, not as something thrust into the system, which requires to be expelled by force, but as an aberration from a natural mode of action, produced by some external cause, we should be in less haste to attack it by medicine, and more watchful in its prevention. Accordingly, where a constant demand for medicine exists in a nursery, the mother may rest assured, that there is something essentially wrong in the treatment of her children.

"Much havoc is made among infants, by the abuse of calomel and other medicines, which procure momentary relief, but end by producing incurable disease; and it has often excited my astonishment, to see how recklessly remedies of this kind are had recourse to, on the most trifling occasions, by mothers and nurses, who would be horrified, if they knew the nature of the power they are wielding, and the extent of injury they are inflicting."

Instead, then, of depending on medicine, for the preservation of the health and life of an infant, the following precautions and preventives should be adopted.

Take particular care of the *food* of an infant. If it is nourished by the mother, her own diet should be simple, nourishing, and temperate. If the child be brought up by hand, the milk of a new-milch cow, mixed with one third water, and sweetened a little with *white* sugar, should be the only food given, until the teeth come. This is more suitable, than any preparations of flour or arrow-root, the nourishment of which is too highly concentrated. Never give a child *bread*, *cake*, or *meat*, before the teeth appear. If the food appear to distress the child, after eating, first ascertain if[Pg 217] the milk be really from a new-milch cow, as it may otherwise be too old. Learn, also, whether the cow lives on proper food. Cows that are fed on *still-slops*, as is often the case in cities, furnish milk which is very unhealthful.

Be sure and keep a good supply of pure and fresh air, in the nursery. On this point, Dr. Bell remarks, respecting rooms constructed without fireplaces, and without doors or windows to let in pure air, from without, "The sufferings of children of feeble constitutions, are increased, beyond measure, by such lodgings as these. *An action, brought by the Commonwealth,* ought to lie against those persons, who build houses for sale or rent, in which rooms are so constructed as not to allow of free ventilation; and *a writ of lunacy* taken out against those, who, with the common-sense experience which all have on this head, should spend any portion of their time, still more, should sleep, in rooms thus nearly air-tight."

After it is a month or two old, take an infant out to walk, or ride, in a little wagon, every fair and warm day; but be very careful that its feet, and every part of its body, are kept warm: and be sure that its eyes are well protected from the light. Weak eyes, and sometimes blindness, are caused by neglecting this precaution. Keep the head of an infant cool, never allowing too warm bonnets, nor permitting it to sink into soft pillows, when asleep. Keeping an infant's head too warm, very much increases nervous irritability; and this is the reason why medical men forbid the use of caps for infants. But the head of an infant should, especially while sleeping, be protected from draughts of air, and from getting cold.

Be very careful of the skin of an infant, as nothing tends so effectually to prevent disease. For this end, it should be washed all over, every morning, and then gentle friction should be applied, with the hand, to the back, stomach, bowels, and limbs. The head should be thoroughly washed, every day, and then brushed[Pg 218] with a soft hair-brush, or combed with a fine comb. If, by neglect, dirt accumulates under the hair, apply, with the finger, the yolk of an egg, and then the fine comb will remove it all, without any trouble.

Dress the infant, so that it will be always warm, but not so as to cause perspiration. Be sure and keep its feet *always* warm; and, for this end, often warm them at a fire, and use long dresses. Keep the neck and arms covered. For this purpose, wrappers, open in front, made high in the neck, with long sleeves, to put on over the frock, are now very fashionable.

It is better for both mother and child, that it should not sleep on the mother's arm, at night, unless the weather be extremely cold. This practice keeps the child too warm, and leads it to seek food too frequently. A child should ordinarily take nourishment but twice in the night. A crib beside the mother, with a plenty of warm and light covering, is best for the child; but the mother must be sure that it is always kept warm. Never cover a child's head, so that it will inhale the air of its own lungs. In very warm weather, especially in cities, great pains should be taken, to find fresh and cool air, by rides and sailing. Walks in a public square, in the cool of the morning, and frequent excursions in ferry or steam-boats, would often save a long bill for medical attendance. In hot nights, the windows should be kept open, and the infant laid on a mattress, or on folded blankets. A bit of straw matting, laid over a featherbed, and covered with the under sheet, makes a very cool bed for an infant.

Cool bathing, in hot weather, is very useful; but the water should be very little cooler than the skin of the child. When the constitution is delicate, the water should be slightly warmed. Simply sponging the body, freely, in a tub, answers the same purpose as a regular bath. In very warm weather, this should be done two or three times a day, always waiting two or three hours after food has been given.

"When the stomach is peculiarly irritable, (from[Pg 219] teething,) it is of paramount necessity to withhold all the nostrums which have been so falsely lauded as 'sovereign cures for *cholera infantum*.' The true restoratives, to a child threatened with disease, are, cool air, cool bathing, and cool drinks of simple water, in addition to *proper* food, at stated intervals." Do not take the advice of mothers, who tell of this, that, and the other thing, which have proved excellent remedies in their experience. Children have different constitutions, and there are multitudes of different causes for their sickness; and what might cure one child, might kill another, which *appeared* to have the same complaint. A mother should go on the general rule, of giving an infant very little medicine, and then only by the direction of a discreet and experienced physician. And there are cases, when, according to the views of the most distinguished and competent practitioners, physicians themselves are much too free in using medicines, instead of adopting *preventive* measures.

Do not allow a child to form such habits, that it will not be quiet, unless tended and amused. A healthy child should be accustomed to lie or sit in its cradle, much of the time; but it should occasionally be taken up, and tossed, or carried about, for exercise and amusement. An infant should be encouraged to *creep*, as an exercise very strengthening and useful. If the mother fears the soiling of its nice dresses, she can keep a long slip or apron, which will entirely cover the dress, and can be removed, when the child is taken in the arms. A child should not be allowed, when quite young, to bear its weight on its feet, very long at a time, as this tends to weaken and distort the limbs.

Many mothers, with a little painstaking, succeed in putting their infants, while awake, into their cradle, at regular hours, for sleep, and induce regularity in other habits, which saves much trouble. In doing this, a child may cry, at first, a great deal; but for a healthy child, this use of the lungs does no harm, and tends rather to strengthen, than to injure, them. A child who[Pg 220] is trained to lie or sit, and amuse itself, is happier than one who is carried and tended a great deal, and thus rendered restless and uneasy when not so indulged.

CHAPTER XX.
ON THE MANAGEMENT OF YOUNG CHILDREN.

In regard to the physical education of children, Dr. Clarke, Physician in Ordinary to the Queen of England, expresses views, on one point, in which most physicians would coincide. He says, "There is no greater error in the management of children, than that of giving them animal diet very early. By persevering in the use of an over-stimulating diet, the digestive organs become irritated, and the various secretions, immediately connected with, and necessary to, digestion, are diminished, especially the *biliary secretion*. Children, so fed, become very liable to attacks of fever, and of inflammation, affecting, particularly, the mucous membranes; and measles, and the other diseases incident to childhood, are generally severe in their attack."

There are some popular notions on the subject of the use of animal food, which need to be corrected.

One mistake, is, in supposing that the formation of the human teeth and stomach indicate that man was designed to feed on flesh. Linnæus says, that the organization of man, when compared with other animals, shows, that "fruits and esculent vegetables constitute his most suitable food." Baron Cuvier, the highest authority on comparative anatomy, says, "the natural food of man, *judging from his structure*, appears to consist of fruits, roots, and other succulent parts of vegetables."

Another common mistake, is, that the stimulus of animal food is necessary for the full developement of the physical and intellectual powers. This notion is disproved by facts. The inhabitants of Lapland and[Pg 221] Kamtschatka, who live altogether on animal food, are among the smallest, weakest, and most timid, of races. But the Scotch Highlanders, who, in a very cold climate, live almost exclusively on milk and vegetable diet, are among the bravest, largest, and most athletic, of men. The South-Sea Islanders, who live almost exclusively on fruits and vegetables, are said to be altogether superior to English sailors, in strength and agility. An intelligent gentleman, who spent many months in Siberia, testifies, that no exiles endure the climate better than those, who have all their lives been accustomed to a vegetable diet. The stoutest and largest tribes in Africa, live solely on vegetable diet, and the bright, intelligent, and active Arabs, live entirely on milk and vegetables.

The popular notion is, that animal food is more nourishing than vegetable; but on this point, scientific men hold different opinions. Experiments, repeatedly made by some chemists, seem to prove the contrary. Tables have been prepared, showing the amount of nutriment in each kind of food, by which it would appear, that, while beef contains thirty-five per cent. of nutritious matter, wheat-bread and rice contain from eighty to ninety-five per cent. The supposed mistake is attributed to the fact, that, on account of the stimulating nature of animal food, it digests easier and more quickly than vegetables. Many physicians, however, among them, Dr. Combe,[O] are of opinion, that animal food "contains a greater quantity of nutriment in a given bulk, than either herbaceous or farinaceous food." In some diseases, too, meat is better for the stomach than vegetables.

The largest proportion of those, who have been remarkable for having lived to the greatest age, were persons, whose diet was almost exclusively vegetables; and it is a well-known fact, that the pulse of a hardy and robust man, who lives on simple vegetable diet, is from [Pg 222]ten to twenty beats less in a minute, than that of men who live on a mixed diet.

In regard to the intellect, Dr. Franklin asserted, from experience, that an exclusively vegetable diet "promotes clearness of ideas and quickness of perception; and is to be preferred, by all who labor with the mind." The mightiest efforts of Sir Isaac Newton, were performed, while nourished only by bread and water. Many other men, distinguished by intellectual vigor, give similar testimony. These facts show that animal food is not needful, to secure the perfect developement of mind or body.[P]

The result of the treatment of the inmates of the Orphan Asylum, at Albany, is one, upon which all, who have the care of young children, should deeply ponder. During the first six years of the existence of this Institution, its average number of children was eighty. For the first three years, their diet was meat once a day, fine bread, rice, Indian puddings, vegetables, fruit, and milk. Considerable attention was given to clothing, fresh air, and exercise; and they were bathed once in three weeks. During these three years, from four to six children, and sometimes more, were continually on the sick-list; one or two assistant nurses were necessary; a physician was called, two or three times a week; and, in this time, there were between thirty and forty deaths. At the end of this period, the management was changed, in these respects:—daily ablutions of the whole body were practised; bread of unbolted flour was substituted for that of fine wheat; and all animal food was banished. More attention also was paid to clothing, bedding, fresh air, and exercise. The result was, that the nursery was vacated; the nurse and physician were no longer needed; and, for two years, not a single case of sickness or death occurred. The [Pg 223]third year, also, there were no deaths, except those of two idiots and one other child, all of whom were new inmates, who had not been subjected to this treatment. The teachers of the children also testified, that there was a manifest increase of intellectual vigor and activity, while there was much less irritability of temper.

Let parents, nurses, and teachers, reflect on the above statement, and bear in mind, that stupidity of intellect, and irritability of temper, as well as ill health, are often caused by the mismanagement of the nursery, in regard to the physical training of children. There is probably no practice, more deleterious, than that of allowing children to eat at short intervals, through the day. As the stomach is thus kept constantly at work, with no time for repose, its functions are deranged, and a weak or disordered stomach is the frequent result. Children should be required to keep cakes, nuts, and other good things which they may have to eat, till just before a meal, and then they will form a part of their regular supply. This is better, than to wait till after their hunger is satisfied by food, when they will eat their niceties merely to gratify the palate, and thus overload the stomach.

In regard to the intellectual training of young children, some modification in the common practice is necessary, with reference to their physical wellbeing. More care is needful, in providing *well-ventilated* schoolrooms, and in securing more time for sports in the open air, during school hours. It is very important, to most mothers, that their young children should be removed from their care, during the six school hours; and it is very useful, to quite young children, to be subjected to the discipline of a school, and to intercourse with other children of their own age. And, with a suitable teacher, it is no matter how early children are sent to school, provided their health is not endangered, by impure air, too much confinement, and too great mental stimulus.

In regard to the formation of the moral character, it[Pg 224] has been too much the case, that the discipline of the nursery has consisted of disconnected efforts to make children either do, or refrain from doing, certain particular acts. Do this, and be rewarded; do that, and be punished; is the ordinary routine of family government.

But children can be very early taught, that their happiness, both now and hereafter, depends on the formation of *habits of submission, self-denial*, and *benevolence*. And all the discipline of the nursery can be conducted by the parents, not only with this general aim in their own minds, but also with the same object daily set before the minds of the children. Whenever their wishes are crossed, or their wills subdued, they can be taught, that all this is done, not merely to please the parent, or to secure some good to themselves or to others; but as a part of that merciful training, which is designed to form such a character, and such habits, that they can hereafter find their chief happiness in giving up their will to God, and in living to do good to others, instead of living merely to please themselves.

It can be pointed out to them, that they must always submit their will to the will of God, or else be continually miserable. It can be shown, how in the nursery, and in the school, and through all future days, a child must practise the giving up of his will and wishes, when they interfere with the rights and comfort of others; and how important it is, early to learn to do this, so that it will, by habit, become easy and agreeable. It can be shown, how children, who are indulged in all their wishes, and who are never accustomed to any self-denial, always find it hard to refrain from what injures themselves and others. It can be shown, also, how important it is, for every person, to form such habits of benevolence, towards others, that self-denial, in doing good, will become easy.

Parents have learned, by experience, that children can be constrained, by authority and penalties, to exercise self-denial, for *their own* good, till a habit is formed,[Pg 225] which makes the duty comparatively easy. For example, well-trained children can be accustomed to deny themselves tempting articles of food, which are injurious, until the practice ceases to be painful and difficult. Whereas, an indulged child would be thrown into fits of anger or discontent, when its wishes were crossed, by restraints of this kind.

But it has not been so readily discerned, that the same method is needful, in order to form a habit of self-denial, in doing good to others. It has been supposed, that, while children must be forced, by *authority*, to be self-denying and prudent, in regard to their own happiness, it may properly be left to their own discretion, whether they will practise any self-denial in doing good to others. But the more difficult a duty is, the greater is the need of parental authority, in forming a habit, which will make that duty easy.

In order to secure this, some parents turn their earliest efforts to this object. They require the young child always to offer to others a part of every thing which it receives; always to comply with all reasonable requests of others for service; and often to practise little acts of self-denial, in order to secure some enjoyment for others. If one child receives a present of some nicety, he is required to share it with all his brothers and sisters. If one asks his brother to help him in some sport, and is met with a denial, the parent requires the unwilling child to act benevolently, and give up some of his time to increase his brother's enjoyment. Of course, in such an effort as this, discretion must be used, as to the frequency and extent of the exercise of authority, to induce a habit of benevolence. But, where parents deliberately aim at such an object, and wisely conduct their instructions and discipline to secure it, very much will be accomplished.

Religious influence should be brought to bear directly upon this point. In the very beginning of religious instruction, Jesus Christ should be presented to the child, as that great and good Being, who came into this[Pg 226] world to teach children how to be happy, both here and hereafter. He, who made it His meat and drink to do the will of His Heavenly Father; who, in the humblest station, and most destitute condition, denied Himself, daily, and went about doing good; should constantly be presented as the object of their imitation. And as nothing so strongly influences the minds of children, as the sympathy and example of a *present* friend, all those, who believe Him to be an *ever-present Saviour*, should avail themselves of this powerful aid. Under such training, Jesus Christ should be constantly presented to them, as their ever-watchful, tender, and sympathizing friend. If the abstract idea of an unembodied Spirit with the majestic attributes of Deity, be difficult for the mind of infancy to grasp, the simple, the gentle, the lovely, character of Christ, is exactly adapted to the wants and comprehension of a child. In this view, how touching is the language of the Saviour, to His misjudging disciples, "Suffer *the little children* to come unto me!"

In regard to forming habits of obedience, there have been two extremes, both of which need to be shunned. One is, a stern and unsympathizing maintenance of parental authority, demanding perfect and constant obedience, without any attempt to convince a child of the propriety and benevolence of the requisitions, and without any manifestation of sympathy and tenderness for the pain and difficulties which are to be met. Under such discipline, children grow up to fear their parents, rather than to love and trust them; while some of the most valuable principles of character, are chilled, or forever blasted.

In shunning this danger, other parents pass to the opposite extreme. They put themselves too much on the footing of equals with their children, as if little were due to superiority of relation, age, and experience. Nothing is exacted, without the implied concession that the child is to be a judge of the propriety of the[Pg 227] requisition; and reason and persuasion are employed, where simple command and obedience would be far better. This system produces a most pernicious influence. Children soon perceive the position, thus allowed them, and take every advantage of it. They soon learn to dispute parental requirements, acquire habits of forwardness and conceit, assume disrespectful manners and address, maintain their views with pertinacity, and yield to authority with ill-humor and resentment, as if their rights were infringed.

The medium course, is, for the parent to take the attitude of a superior, in age, knowledge, and relation, who has a perfect right to control every action of the child, and that, too, without giving any reason for the requisitions. "Obey, *because your parent commands*," is always a proper and sufficient reason.

But care should be taken, to convince the child that the parent is conducting a course of discipline, designed to make him happy; and in forming habits of implicit obedience, self-denial, and benevolence, the child should have the reasons for most requisitions kindly stated; never, however, on the demand of it, from the child, as a right, but as an act of kindness from the parent.

It is impossible to govern children properly, especially those of strong and sensitive feelings, without a constant effort to appreciate the value which they attach to their enjoyments and pursuits. A lady, of great strength of mind and sensibility, once told the writer, that one of the most acute periods of suffering, in her whole life, was occasioned by the burning up of some milkweed-silk, by her mother. The child had found, for the first time, some of this shining and beautiful substance; was filled with delight at her discovery; was arranging it in parcels; planning its future uses, and her pleasure in showing it to her companions,—when her mother, finding it strewed over the carpet, hastily swept it into the fire, and that, too, with so indifferent an air, that the child fled away, almost distracted [Pg 228]with grief and disappointment. The mother little realized the pain she had inflicted, but the child felt the unkindness, so severely, that for several days her mother was an object almost of aversion.

While, therefore, the parent needs to carry on a steady course, which will oblige the child always to give up its will, whenever its own good, or the greater claims of others, require it, this should be constantly connected with the expression of a tender sympathy, for the trials and disappointments thus inflicted. Those, who will join with children, and help them along in their sports, will learn, by this mode, to understand the feelings and interests of childhood; while, at the same time, they secure a degree of confidence and affection, which cannot be gained so easily, in any other way. And it is to be regretted, that parents so often relinquish this most powerful mode of influence, to domestics and playmates, who often use it in the most pernicious manner. In joining in such sports, older persons should never relinquish the attitude of superiors, or allow disrespectful manners or address. And respectful deportment is never more cheerfully accorded, than in seasons, when young hearts are pleased, and made grateful, by having their tastes and enjoyments so efficiently promoted.

Next to the want of all government, the two most fruitful sources of evil to children, are, *unsteadiness* in government, and *over-government*. Most of the cases, in which the children of sensible and conscientious parents turn out badly, result from one or the other of these causes. In cases of unsteady government, either one parent is very strict, severe, and unbending, and the other excessively indulgent, or else the parents are sometimes very strict and decided, and at other times allow disobedience to go unpunished. In such cases, children, never knowing exactly when they can escape with impunity, are constantly tempted to make the trial.

The bad effects of this, can be better appreciated, by[Pg 229] reference to one important principle of the mind. It is found to be universally true, that, when any object of desire is put entirely beyond the reach of hope or expectation, the mind very soon ceases to long for it, and turns to other objects of pursuit. But, so long as the mind is hoping for some good, and making efforts to obtain it, any opposition excites irritable feelings. Let the object be put entirely beyond all hope, and this irritation soon ceases. In consequence of this principle, those children, who are under the care of persons of steady and decided government, know, that whenever a thing is forbidden or denied, it is out of the reach of hope; the desire, therefore, soon ceases, and they turn to other objects. But the children of undecided, or of over-indulgent parents, never enjoy this preserving aid. When a thing is denied, they never know but either coaxing may win it, or disobedience secure it without any penalty, and so they are kept in that state of hope and anxiety, which produces irritation, and tempts to insubordination. The children of very indulgent parents, and of those who are undecided and unsteady in government, are very apt to become fretful, irritable, and fractious.

Another class of persons, in shunning this evil, go to the other extreme, and are very strict and pertinacious, in regard to every requisition. With them, fault-finding and penalties abound, until the children are either hardened into indifference of feeling, and obtuseness of conscience, or else become excessively irritable, or misanthropic.

It demands great wisdom, patience, and self-control, to escape these two extremes. In aiming at this, there are parents, who have found the following maxims of very great value. First, Avoid, as much as possible, the multiplication of rules and absolute commands. Instead of this, take the attitude of advisers. "My child, this is improper, I wish you would remember not to do it." This mode of address answers for all the little acts of heedlessness, awkwardness, or ill-manners,[Pg 230] so frequently occurring, with children. There are cases, when direct and distinct commands are needful; and, in such cases, a penalty for disobedience should be as steady and sure as the laws of Nature. Where such steadiness, and certainty of penalty, attend disobedience, children no more think of disobeying, than they do of putting their fingers in a burning candle.

The next maxim, is, Govern by rewards, more than by penalties. Such faults as wilful disobedience, lying, dishonesty, and indecent or profane language, should be punished with severe penalties, after a child has been fully instructed in the evil of such practices. But all the constantly-recurring faults of the nursery, such as ill-humor, quarrelling, carelessness, and ill-manners, may, in a great many cases, be regulated by gentle and kind remonstrances, and by the offer of some reward for persevering efforts to form a good habit. It is very injurious and degrading to any mind, to be kept under the constant fear of penalties. *Love* and *hope* are the principles that should be mainly relied on, in forming the habits of childhood.

Another maxim, and perhaps the most difficult, is, Do not govern by the aid of severe and angry tones. A single example will be given to illustrate this maxim. A child is disposed to talk and amuse itself, at table. The mother requests it to be silent, except when needing to ask for food, or when spoken to by its older friends. It constantly forgets. The mother, instead of rebuking, in an impatient tone, says, "My child, you must remember not to talk. I will remind you of it four times more, and after that, whenever you forget, you must leave the table, and wait till we are done." If the mother is steady in her government, it is not probable that she will have to apply this slight penalty more than once or twice. This method is far more effectual, than the use of sharp and severe tones, to secure attention and recollection, and often answers the purpose, as well as offering some reward.

The writer has been in some families, where the[Pg 231] most efficient and steady government has been sustained, without the use of a cross or angry tone; and in others, where a far less efficient discipline was kept up, by frequent severe rebukes and angry remonstrances. In the first case, the children followed the example set them, and seldom used severe tones to each other; in the latter, the method employed by the parents, was imitated by the children; and cross words and angry tones resounded from morning till night, in every portion of the household.

Another important maxim, is, Try to keep children in a happy state of mind. Every one knows, by experience, that it is easier to do right, and submit to rule, when cheerful and happy, than when irritated. This is peculiarly true of children; and a wise mother, when she finds her child fretful and impatient, and thus constantly doing wrong, will often remedy the whole difficulty, by telling some amusing story, or by getting the child engaged in some amusing sport. This strongly shows the importance of learning to govern children without the employment of angry tones, which always produce irritation.

Children of active, heedless temperament, or those who are odd, awkward, or unsuitable, in their remarks and deportment, are often essentially injured, by a want of patience and self-control in those who govern them. Such children, often possess a morbid sensibility, which they strive to conceal, or a desire of love and approbation, which preys like a famine on the soul. And yet, they become objects of ridicule and rebuke, to almost every member of the family, until their sensibilities are tortured into obtuseness or misanthropy. Such children, above all others, need tenderness and sympathy. A thousand instances of mistake or forgetfulness should be passed over, in silence, while opportunities for commendation and encouragement should be diligently sought.

In regard to the formation of habits of self-denial, in childhood, it is astonishing to see how parents, who are[Pg 232] very sensible, often seem to regard this matter. Instead of inuring their children to this duty, in early life, so that by habit it may be made easy in after-days, they seem to be studiously seeking to cut them off, from every chance to secure such a preparation. Every wish of the child is studiously gratified; and, where a necessity exists, of crossing its wishes, some compensating pleasure is offered, in return. Such parents, often maintain that nothing shall be put on their table, which their children may not join them in eating. But where, so easily and surely as at the daily meal, can that habit of self-denial be formed, which is so needful in governing the appetites, and which children must acquire, or be ruined? The food which is proper for grown persons, is often unsuitable for children; and this is a sufficient reason for accustoming them to see others partake of delicacies, which they must not share. Requiring children to wait till others are helped, and to refrain from conversation at table, except when addressed by their elders, is another mode of forming habits of self-denial and self-control. Requiring them to help others, first, and to offer the best to others, has a similar influence.

In forming the moral habits of children, it is wise to take into account the peculiar temptations to which they are to be exposed. The people of this Nation are eminently a trafficking people; and the present standard of honesty, as to trade and debts, is very low, and every year seems sinking still lower. It is, therefore, pre-eminently important, that children should be trained to strict *honesty*, both in word and deed. It is not merely teaching children to avoid absolute lying, which is needed. *All kinds of deceit* should be guarded against; and all kinds of little dishonest practices be strenuously opposed. A child should be brought up with the determined principle, never to *run in debt*, but to be content to live in an humbler way, in order to secure that true independence, which should be the noblest distinction of an American citizen.

[Pg 233]

There is no more important duty, devolving upon a mother, than the cultivation of habits of modesty and propriety in young children. All indecorous words or deportment, should be carefully restrained; and delicacy and reserve studiously cherished. It is a common notion, that it is important to secure these virtues to one sex, more than to the other; and, by a strange inconsistency, the sex most exposed to danger, is the one selected as least needing care. But a wise mother will be especially careful, that her sons are trained to modesty and purity of mind.

But few mothers are sufficiently aware of the dreadful penalties which often result from indulged impurity of thought. If children, in *future* life, can be preserved from licentious associates, it is supposed that their safety is secured. But the records of our insane retreats, and the pages of medical writers, teach, that even in solitude, and without being aware of the sin or the danger, children may inflict evils on themselves, which not unfrequently terminate in disease, delirium, and death. Every mother and every teacher, therefore, carefully avoiding all explanation of the mystery, should teach the young, that the indulgence of impure thoughts and actions, is visited by the most awful and terrific penalties. Disclosing the details of vice, in order to awaken dread of its penalties, is a most dangerous experiment, and often leads to the very evils feared. The attempts made, in late years, to guard children from future dangers, by circulating papers, and books of warning and information, have led to such frightful results, that it is hoped the experiment will never again be pursued. The safest course, is, to cultivate habits of modesty and delicacy, and to teach, that all impure thoughts, words, and actions, are forbidden by God, and are often visited by the most dreadful punishment. At the same time, it is important for mothers to protect the young mind from false notions of delicacy. It should be shown, that whatever is necessary, to save from suffering or danger, must be met, without shame or aversion; and[Pg 234] that all, which God has instituted, is wise, and right, and pure.

It is in reference to these dangers, that mothers and teachers should carefully guard the young from those highly-wrought fictions, which lead the imagination astray; and especially from that class of licentious works, made interesting by genius and taste, which have flooded this Country, and which are often found on the parlor table, even of moral and Christian people. Of this class, the writings of Bulwer stand conspicuous. The only difference, between some of his works and the obscene prints, for vending which men suffer the penalties of the law, is, that the last are so gross, as to revolt the taste and startle the mind to resistance, while Bulwer presents the same ideas, so clothed in the fascinations of taste and genius, as most insidiously to seduce the unwary. It seems to be the chief aim of this licentious writer, to make thieves, murderers, and adulterers, appear beautiful, refined, and interesting. It is time that all virtuous persons in the community should rise in indignation, not only against the writers, but the venders of such poison.

FOOTNOTES:

[O] See his 'Physiology of Digestion considered with relation to the Principles of Dietetics,' issued by the Publishers of this work.

[P] The writer is not an advocate for *total* abstinence from animal food. She coincides with the best authorities, in thinking that adults eat too much; that children, while growing, should eat very little, and quite young children, none at all.

CHAPTER XXI.
ON THE CARE OF THE SICK.

Every woman who has the care of young children, or of a large family, is frequently called upon, to advise what shall be done, for some one who is indisposed; and often, in circumstances where she must trust solely to her own judgement. In such cases, some err, by neglecting to do any thing at all, till the patient is quite sick; but a still greater number err, from excessive and injurious dosing.

The two great causes of the ordinary slight attacks of illness, in a family, are, sudden chills, which close[Pg 235] the pores of the skin, and thus affect the throat, lungs, or bowels; and the excessive or improper use of food. In most cases, of illness from the first cause, bathing the feet, and some aperient drink to induce perspiration, are suitable remedies. A slight cathartic, also, is often serviceable. In case of illness from improper food, or excess in eating, *fasting*, for one or two meals, to give the system time and chance to relieve itself, is the safest remedy. Sometimes, a gentle cathartic may be needful; but it is best first to try fasting.

The following extract from a discourse of Dr. Burne, before the London Medical Society, contains important information. "In civilized life, the causes, which are most generally and continually operating in the production of diseases, are, affections of the mind, improper diet, and retention of the intestinal excretions. The undue retention of excrementitious matter, allows of the absorption of its more liquid parts, which is a cause of great impurity to the blood, and the excretions, thus rendered hard and knotty, act more or less as extraneous substances, and, by their irritation, produce a determination of blood to the intestines and to the neighboring viscera, which ultimately ends in inflammation. It also has a great effect on the whole system; causes a determination of blood to the head, which oppresses the brain and dejects the mind; deranges the functions of the stomach; causes flatulency; and produces a general state of discomfort."

Dr. Combe remarks, on this subject, "In the natural and healthy state, under a proper system of diet, and with sufficient exercise, the bowels are relieved regularly, once every day." *Habit* "is powerful in modifying the result, and in sustaining healthy action when once fairly established. Hence the obvious advantage of observing as much regularity, in relieving the system, as in taking our meals." It is often the case that soliciting Nature at a regular period, once a day, will remedy constipation, without medicine, and induce a regular and healthy state of the bowels. "When,[Pg 236] however, as most frequently happens, the constipation arises from the absence of all assistance from the abdominal and respiratory muscles, the first step to be taken, is, again to solicit their aid; first, by removing all impediments to free respiration, such as stays, waistbands and belts; secondly, by resorting to such active exercises, as shall call the muscles into full and regular action; and, lastly, by proportioning the quantity of food to the wants of the system, and the condition of the digestive organs. If we employ these means, systematically and perseveringly, we shall rarely fail in at last restoring the healthy action of the bowels, with little aid from medicine. But if we neglect these modes, we may go on, for years, adding pill to pill, and dose to dose, without ever attaining the end at which we aim." There is no point, in which a woman needs more knowledge and discretion, than in administering remedies for what seem slight attacks, which are not supposed to require the attention of a physician. It is little realized, that purgative drugs are unnatural modes of stimulating the internal organs, tending to exhaust them of their secretions, and to debilitate and disturb the animal economy. For this reason, they should be used as little as possible; and fasting, and perspiration, and the other methods pointed out, should always be first resorted to. When medicine must be given, it should be borne in mind, that there are various classes of purgatives, which produce very diverse effects. Some, like salts, operate to thin the blood, and reduce the system; others are stimulating; and others have a peculiar operation on certain organs. Of course, great discrimination and knowledge is needed, in order to select the kind, which is suitable to the particular disease, or to the particular constitution of the invalid. This shows the folly of using the many kinds of pills, and other quack medicines, where no knowledge can be had of their composition. Pills which are good for one kind of disease, might operate as poison in another state of the system. It is wise to keep always on hand[Pg 237] some simple cathartic, for family use, in slight attacks; and always to resort to medical advice, whenever powerful remedies seem to be demanded.[Q] It is very common, in cases of colds which affect the lungs or throat, to continue to try one dose after another, for relief. It will be well to bear in mind, at such times, that all which goes into the stomach, must be first absorbed into the blood, before it can reach the diseased part; and that there is some danger of injuring the stomach, or other parts of the system, by such a variety of doses, many of which, it is probable, will be directly contradictory in their nature, and thus neutralize any supposed benefit they might separately impart.

It is very unwise, to tempt the appetite of a person who is indisposed. The cessation of appetite is the warning of Nature, that the system is in such a state, that food cannot be digested.

The following suggestions may be found useful, in regard to nursing the sick. As nothing contributes more to the restoration of health, than pure air, it should be a primary object, to keep a sick-room well ventilated. At least twice in the twenty-four hours, the patient should be well covered, and fresh air freely admitted from out of doors. After this, if need be, the room should be restored to a proper temperature, by the aid of a fire. Bedding and clothing should also be well aired, and frequently changed; as the exhalations from the body, in sickness, are peculiarly deleterious. Frequent ablutions, of the whole body, if possible, are very useful; and for these, warm water may be employed.

[Pg 238]

The following, are useful directions for dressing a blister. Spread thinly, on a linen cloth, an ointment, composed of one third of beeswax to two thirds of tallow; lay this upon a linen cloth, folded many times. With a sharp pair of scissors, make an aperture in the lower part of the bag of water, with a little hole, above, to give it vent. Break the raised skin as little as possible. Lay on the cloth, spread as directed. The blister, at first, should be dressed as often as three times in a day, and the dressing renewed each time.

A sick-room should always be kept very neat, and in perfect order; and all haste, noise, and bustle, should be avoided. In order to secure neatness, order, and quiet, in case of long illness, the following arrangements should be made. Keep a large box for fuel, which will need to be filled only twice in twenty-four hours. Provide, also, and keep in the room, or an adjacent closet, a small teakettle, a saucepan, a pail of water, for drinks and ablutions, a pitcher, a covered porringer, two pint bowls, two tumblers, two cups and saucers, two wine glasses, two large and two small spoons; also, a dish in which to wash these articles; a good supply of towels, and a broom. Keep a slop-bucket, near by, to receive the wash of the room. Procuring all these articles at once, will save much noise and confusion.

Whenever medicine or food is given, spread a clean towel over the person or bedclothing, and get a clean handkerchief, as nothing is more annoying to a weak stomach, than the stickiness and soiling produced by medicine and food. Keep the fireplace neat, and always wash all articles, and put them in order, as soon as they are out of use.

A sick person has nothing to do, but look about the room; and when every thing is neat and in order, a feeling of comfort is induced, while disorder, filth, and neglect, are constant objects of annoyance, which, if not complained of, are yet felt.

[Pg 239]

Always prepare food for the sick, in the neatest and most careful manner. It is in sickness, that the senses of smell and taste are most susceptible of annoyance; and often, little mistakes or negligences, in preparing food, will take away all appetite.

Food for the sick, should be cooked on coals, that no smoke may have access to it; and great care must be taken, to prevent any adherence to the bottom, as this always gives a disagreeable taste.

Keeping clean handkerchiefs and towels at hand, cooling the pillows, sponging the hands with water, swabbing the mouth with a clean linen rag, on the end of a stick, are modes of increasing the comfort of the sick. Always throw a shawl over a sick person, when raised up.

Be careful to understand a physician's directions, and *to obey them implicitly*. If it be supposed that any other person knows better about the case, than the physician, dismiss the physician, and employ that person in his stead.

In nursing the sick, always speak gently and cheeringly; and, while you express sympathy for their pain and trials, stimulate them to bear all with fortitude, and with resignation to Him who has appointed the trial. Offer to read the Bible, or other devotional books, whenever it is suitable, and will not be deemed obtrusive.

It is always best to consult the physician, as to where medicines shall be purchased, and to show the articles to him before using them, as great impositions are practised in selling old, useless, and adulterated drugs. Always put labels on vials of medicine, and keep them out of the reach of children.

Be careful to label all powders, and particularly all *white powders*; as many poisonous medicines, in this form, are easily mistaken for others which are harmless.

FOOTNOTE:

[Q] The following electuary, by a distinguished physician, is used by many friends of the writer, as a standing resort, in cases of constipation, or where a gentle cathartic is needed. One recommendation of it, is, that children always love it, and eat the pills as "good plums."

Two ounces of powdered Senna; one ounce of Cream of Tartar; one ounce of Sulphur; mixed with sufficient Confection of Senna, to form an electuary. Make this into pills, of the size of peas, and give a young child two or three, as the case may be. Taking three pills, every night, will generally relieve constipation in an adult.

[Pg 240]

CHAPTER XXII.
ON ACCIDENTS AND ANTIDOTES.

When serious accidents occur, medical aid should be immediately procured. Till that can be done, the following directions may be useful.

When a child has any thing in its throat, first try, with the finger, to get the article up. If this cannot be done, push it down into the stomach, with a smooth elastic stick. If the article be a pin, sharp bone, glass, or other cutting substance, give an emetic which will immediately operate.

In the case of a common cut, bind the lips of the wound together, with a rag, and put nothing else on. If the cut be large, and so situated that rags will not bind it together, use sticking plaster, cut in strips and laid obliquely across the cut. Sometimes it is needful to take a stitch, with a needle and thread, on each lip of the wound, and draw the two sides together.

If an artery be cut, it must be immediately tied up, or the person will bleed to death. The blood from an artery is of a bright red color, and spirts out, in regular jets, at each beat of the heart. Take up the bleeding end of the artery, and hold it, or tie it up, till a surgeon comes. When the artery cannot be found, and in all cases of bad cuts on any of the limbs, apply compression; when it can be done, tie a very tight bandage above the wound, if it be below the heart, and *below* if the wound be above the heart. Put a stick into the band, and twist it as tight as can be borne, till surgical aid be obtained.

Bathe bad bruises in hot water, or hot spirits, or a decoction of bitter herbs. *Entire rest,* is the remedy for sprains. Bathing in warm water, or warm whiskey is very useful. A sprained leg should be kept in a horizontal position, on a bed or sofa.

When a leg is broken, tie it to the other leg, to keep[Pg 241] it still; and, if possible, get a surgeon, before the limb swells. Bind a broken arm to a piece of shingle, and keep it still, till it is set.

In case of a blow on the head, or a fall, causing insensibility, use a mustard paste on the back of the neck and pit of the stomach, and rub the body with spirits. After the circulation is restored, bleeding is often necessary; but it is very dangerous to attempt it before.

In cases of bad burns, where the skin is taken off, the great aim should be, *to keep the injured part from the air.* For this purpose, sprinkle on flour, or apply a liniment, made of linseed oil and lime-water, in equal quantities. Sweet-oil, on cotton, is good, and with laudanum, alleviates pain: but many skins cannot bear the application of raw cotton, which is sometimes very good. When a dressing is put on, do not remove it, as it will be sure to protract the cure, by admitting the air.

In case of drowning, lay the person in a warm bed, or on blankets, on the right side, with the head raised, and a little inclined forward. Clear the mouth with the fingers, and cautiously apply hartshorn to the nose. Raise the heat of the body, by bottles of warm water, applied to the pit of the stomach, armpits, groins, and soles of the feet. Apply friction to the whole body, with warm hands and cloths dipped in warm spirits of camphor. Endeavor to produce the natural action of the lungs, by introducing the nose of a bellows into one nostril and closing the other, at the same time pressing on the throat, to close the gullet. When the lungs are thus inflated, press gently on the breast and belly, and continue the process, for a long time. Cases have been known, where efforts have been protracted eight or ten hours, without effect, and then have proved successful. Rolling the body on a barrel, suspending it by the heels, giving injections of tobacco, and many other practices, which have been common, are highly injurious. After signs of life appear, give small quantities of wine, or spirits and water.

In cases of poisoning, from *corrosive sublimate,* beat[Pg 242] up the whites of twelve eggs, mix them in two quarts of water, and give a tumbler full every three minutes, till vomiting is produced. This is the surest remedy. When this is not at hand, fill the stomach, in like manner, with any mucilaginous substance, such as gum and water, flaxseed, or slippery-elm-bark tea. Flour and water, or sugar and water, in great quantities, are next best; and if none of these be at hand, give copious draughts of water alone.

In case of poisoning from *arsenic, cobalt,* or any such mineral, administer, as soon as possible, large quantities of lime-water and sugared-water, of warm, or even of cold water, or of flaxseed tea, or some other mucilaginous drink, to distend the stomach and produce immediate vomiting, and thereby eject the poison.

If opium, or any of its preparations, has been taken, in dangerous quantities, induce vomiting, without a moment's unnecessary delay, by giving, immediately, in *a small quantity* of water, ten grains of ipecac, and ten grains of sulphate of zinc, (white vitriol, which is the most prompt emetic known,) and repeat the dose every fifteen minutes, till the stomach is entirely emptied. Where white vitriol is not at hand, substitute three or four grains of blue vitriol, (sulphate of copper.) When the stomach is emptied, but not before, give, every ten minutes, alternately, a cup of acid drink, and a cup of very strong coffee, made by pouring a pint of boiling water on a quarter of a pound of ground burnt coffee, and letting it stand ten minutes, and then straining it. Continue these drinks, till the danger is over. Dash cold water on the head, apply friction to the body, and keep the person in constant motion, to prevent sleep.

If any kind of acid be taken, in poisonous quantities, give strong pearlash-water. If ley, or pearlash, or any alkali be taken, give sweet-oil; or, if this be wanting, lamp-oil; or, if neither be at hand, give vinegar, freely.

In case of stupefaction, from the fumes of charcoal, or from entering a well, limekiln, or coal mine, expose[Pg 243] the person to cold air, lying on his back, dash cold water on the head and breast, and rub the body with spirits of camphor, vinegar, or Cologne water. Apply mustard paste to the pit of the stomach, and use friction on the hands, feet, and whole length of the back bone. Give some acid drink, and, when the person revives, place him in a warm bed, in fresh air. Be prompt and persevering.

In case of bleeding at the lungs, or stomach, or throat, give a teaspoonful of dry salt, and repeat it often. For bleeding at the nose, pour cold water on the back of the neck, keeping the head elevated.

If a person be struck with lightning, throw pailfuls of cold water on the head and body, and apply mustard poultices on the stomach, with friction of the whole body, and inflation of the lungs. When no other emetic can be found, pounded mustard seed, taken a teaspoonful at a time, will answer. The ground mustard is not so effectual, but will do.

In case of fire, wrap a woollen blanket about you, to protect from the fire. If the staircases are on fire, tie the corners of the sheets together, very firmly, fasten one end to the bedstead, draw it to the window, and let yourself down. Never read in bed, lest you fall asleep, and the bed be set on fire. If your clothes get on fire, never run, but lie down, and roll about till you can reach a bed or carpet to wrap yourself in, and thus put out the fire. Keep young children in woollen dresses, to save them from the risk of fire.

In thunderstorms, shut the doors and windows. The safest part of a room, is its centre; and where there is a featherbed in the apartment, that will be found the most secure resting-place.

A lightning rod, if it be well pointed, and run deep into the earth, is a certain protection to a circle around it, whose diameter equals the height of the rod above the highest chimney. But it protects *no further* than this extent.

[Pg 244]

CHAPTER XXIII.
ON DOMESTIC AMUSEMENTS AND SOCIAL DUTIES.

Whenever the laws of body and mind are properly understood, it will be allowed, that every person needs some kind of recreation; and that, by seeking it, the body is strengthened, the mind is invigorated, and all our duties are more cheerfully and successfully performed.

Children, whose bodies are rapidly growing, and whose nervous system is tender and excitable, need much more amusement, than persons of mature age. Persons, also, who are oppressed with great responsibilities and duties, or who are taxed by great intellectual or moral excitement, need recreations which secure physical exercise, and draw off the mind from absorbing interests. Unfortunately, such persons are those who least resort to amusements, while the idle, gay, and thoughtless, seek those which are needless, and for which useful occupation would be a most beneficial substitute.

As the only legitimate object of amusements, is, to prepare mind and body for the proper discharge of duty, any protracting of such as interfere with regular employments, or induce excessive fatigue, or weary the mind, or invade the proper hours for repose, must be sinful.

In deciding what should be selected, and what avoided, the following rules are binding. In the first place, no amusements, which inflict needless pain, should ever be allowed. All tricks which cause fright, or vexation, and all sports, which involve suffering to animals, should be utterly forbidden. Hunting and fishing, for mere sport, can never be justified. If a man can convince his children, that he follows these pursuits to gain food or health, and not for amusement, his example may not be very injurious. But, when children see[Pg 245] grown persons kill and frighten animals, for sport, habits of cruelty, rather than feelings of tenderness and benevolence, are induced.

In the next place, we should seek no recreations, which endanger life, or interfere with important duties. As the only legitimate object of amusements, is to promote health, and prepare for more serious duties, selecting those which have a directly opposite tendency, cannot be justified. Of course, if a person feel that the previous day's diversions have shortened the hours of needful repose, or induced a lassitude of mind or body, instead of invigorating them, it is certain that an evil has been done, which should never be repeated.

A third rule, is, to avoid those amusements, which experience has shown to be so exciting, and connected with so many temptations, as to be pernicious in tendency, both to the individual and to the community. It is on this ground, that horse-racing and circus-riding are excluded. Not because there is any thing positively wrong, in having men and horses run, and perform feats of agility, or in persons looking on for the diversion; but because experience has shown so many evils connected with these recreations, that they should be relinquished. So with theatres. The enacting of characters, and the amusement thus afforded, in itself may be harmless; and possibly, in certain cases, might be useful: but experience has shown so many evils to result from this source, that it is deemed wrong to patronize it. So, also, with those exciting games of chance, which are employed in gambling.

Under the same head, comes *dancing*, in the estimation of the great majority of the religious world. Still, there are many intelligent, excellent, and conscientious persons, who hold a contrary opinion. Such maintain, that it is an innocent and healthful amusement, tending to promote ease of manners, cheerfulness, social affection, and health of mind and body; that evils are involved only in its excess; that, like food, study, or religious excitement, it is only wrong, when not properly [Pg 246]regulated; and that, if serious and intelligent people would strive to regulate, rather than banish, this amusement, much more good would be secured.

On the other side, it is objected, not that dancing is a sin, in itself considered, for it was once a part of sacred worship; not that it would be objectionable, if it were properly regulated; not that it does not tend, when used in a proper manner, to health of body and mind, to grace of manners, and to social enjoyment: all these things are conceded. But it is objected to, on the same ground as horse-racing, card-playing, and theatrical entertainments; that we are to look at amusements as they *are*, and not as they *might* be. Horseraces might be so managed, as not to involve cruelty, gambling, drunkenness, and every other vice. And so might theatres and cards. And if serious and intelligent persons, undertook to patronize these, in order to regulate them, perhaps they would be somewhat raised from the depths, to which they are now sunk. But such persons, know, that, with the weak sense of moral obligation existing in the mass of society, and the imperfect ideas mankind have of the proper use of amusements, and the little self-control, which men, or women, or children, practise, these will not, in fact, be thusregulated. And they believe dancing to be liable to the same objections.

As this recreation is actually conducted, it does not tend to produce health of body or mind, but directly the contrary. If young and old went out to dance together, in the open air, as the French peasants do, it would be a very different sort of amusement, from that which is witnessed, in a room, furnished with many lights, and filled with guests, both expending the healthful part of the atmosphere, where the young collect, in their tightest dresses, to protract, for several hours, a kind of physical exertion, which is not habitual to them. During this process, the blood is made to circulate more swiftly than ordinary, in circumstances where it is less perfectly oxygenized than health [Pg 247]requires; the pores of the skin are excited by heat and exercise; the stomach is loaded with indigestible articles, and the quiet, needful to digestion, withheld; the diversion is protracted beyond the usual hour for repose; and then, when the skin is made the most highly susceptible to damps and miasms, the company pass from a warm room to the cold night-air. It is probable, that no single amusement can be pointed out, combining so many injurious particulars, as this, which is so often defended as a healthful one. Even if parents, who train their children to dance, can keep them from public balls, (which is seldom the case,) dancing in private parlors is subject to nearly all the same mischievous influences.

As to the claim of social benefits,—when a dancing-party occupies the parlors, and the music begins, most of the conversation ceases; while the young prepare themselves for future sickness, and the old look smilingly on.

As to the claim for ease and grace of manners,—all that is gained, by this practice, can be better secured, by Calisthenics, which, in all its parts, embraces a much more perfect system, both of healthful exercise, graceful movement, and pleasing carriage.

The writer was once inclined to the common opinion, that dancing was harmless, and might be properly regulated; and she allowed a fair trial to be made, under her auspices, by its advocates. The result was, a full conviction, that it secured no good effect, which could not be better gained another way; that it involved the most pernicious evils to health, character, and happiness; and that those parents were wise, who brought up their children with the full understanding that they were neither to learn nor to practise the art. In the fifteen years, during which she has had the care of young ladies, she has never known any case, where learning this art, and following the amusement, did not have a bad effect, either on the habits, the intellect, the feelings, or the health. Those young ladies, who are[Pg 248] brought up with less exciting recreations, are uniformly likely to be the most contented and most useful, while those, who enter the path to which this diversion leads, acquire a relish and desire for high excitement, which make the more steady and quiet pursuits and enjoyments of home, comparatively tasteless. This, the writer believes to be generally the case, though not invariably so; for there are exceptions to all general rules.

In reference to these exciting amusements, so liable to danger and excess, parents are bound to regard the principle, which is involved in the petition, "Lead us not into temptation." Would it not be inconsistent, to teach this prayer, to the lisping tongue of childhood, and then send it to the dancing-master, to acquire a love for a diversion, which leads to constant temptations that so few find strength to resist?

It is encouraging, to those who take this view of the subject, to find how fast the most serious and intelligent portion of the community is coming to a similar result. Twenty-five years ago, dancing was universally practised by the young, as a matter of course, in every part of the Nation. Now, in those parts of the Country, where religion and intelligence are most extensively diffused, it is almost impossible to get up a ball, among the more refined classes of the community. The amusement is fast leaving this rank in society, to remain as a resource for those, whose grade of intelligence and refinement does not relish more elevated recreations. Still, as there is great diversity of opinion, among persons of equal worth and intelligence, a spirit of candor and courtesy should be practised, on both sides. The sneer at bigotry and narrowness of views, on one side, and the uncharitable implication of want of piety, or sense, on the other, are equally illbred and unchristian. Truth, on this subject, is best promoted, not by ill-natured crimination and rebuke, but by calm reason, generous candor, forbearance, and kindness.

There is another species of amusement, which a[Pg 249] large portion of the religious world have been accustomed to put under the same condemnation as the preceding. This is novel-reading. The confusion and difference of opinion on this subject, have arisen from a want of clear and definite distinctions. Now, as it is impossible to define what are novels and what are not, so as to include one class of fictitious writings and exclude every other, it is impossible to lay down any rule respecting them. The discussion, in fact, turns on the use of those works of imagination, which belong to the class of narratives. That this species of reading, is not only lawful, but necessary and useful, is settled by Divine examples, in the parables and allegories of Scripture. Of course, the question must be, what kind of fabulous writings must be avoided, and what allowed. In deciding this, no specific rules can be given; but it must be a matter to be regulated by the nature and circumstances of each case. No works of fiction, which tend to throw the allurements of taste and genius around vice and crime, should ever be tolerated; and all that tend to give false views of life and duty, should also be banished. Of those, which are written for mere amusement, presenting scenes and events that are interesting and exciting, and having no bad moral influence, much must depend on character and circumstances. Some minds are torpid and phlegmatic, and need to have the imagination stimulated: such would be benefitted by this kind of reading. Others have quick and active imaginations, and would be as much injured. Some persons are often so engaged in absorbing interests, that any thing innocent, which will for a short time draw off the mind, is of the nature of a medicine; and, in such cases, this kind of reading is useful.

There is need, also, that some men should keep a supervision of the current literature of the day, as guardians, to warn others of danger. For this purpose, it is more suitable for *editors, clergymen,* and *teachers,* to [Pg 250]read indiscriminately, than for any other class of persons; for they are the guardians of the public weal, in matters of literature, and should be prepared to advise parents and young persons of the evils in one direction and the good in another. In doing this, however, they are bound to go on the same principles which regulate physicians, when they visit infected districts, — using every precaution to prevent injury to themselves; having as little to do with pernicious exposures, as a benevolent regard to others will allow; and faithfully employing all the knowledge and opportunities, thus gained, for warning and preserving others. There is much danger, in taking this course, that men will seek the excitement of the imagination, for the mere pleasure it affords, under the plea of preparing to serve the public, when this is neither the aim nor the result.

In regard to the use of such works, by the young, as a general rule, they ought not to be allowed to any, except those of a dull and phlegmatic temperament, until the solid parts of education are secured, and a taste for more elevated reading is acquired. If these stimulating condiments in literature be freely used, in youth, all relish for more solid reading, will, in a majority of cases, be destroyed. If parents succeed in securing habits of cheerful and implicit obedience, it will be very easy to regulate this matter, by prohibiting the reading of any story-book, until the consent of the parent is obtained.

It is not unfrequently the case, that advocates for dancing, and the other more exciting amusements, speak as if those, who were more strict in these matters, were aiming to deprive the young of all diversions; just as if, when cards, theatres, and dancing, are cut off, nothing remains but serious and severe duties. Perhaps there has been some just ground of objection to the course often pursued by parents, in neglecting to provide agreeable and suitable substitutes, for the amusements denied; but, there is a great abundance[Pg 251] of safe, healthful, and delightful, recreations, which all parents may secure for their children. Some of these will here be pointed out.

One of the most useful and important, is, the cultivation of flowers and fruits. This, especially for the daughters of a family, is greatly promotive of health and amusement. It is with the hope, that many young ladies, whose habits are now so formed, that they can never be induced to a course of active domestic exercise, so long as their parents are able to hire domestics, may yet be led to an employment, which will tend to secure health and vigor of constitution, that so much space is given, in this work, to directions for the cultivation of fruits and flowers. It would be a most desirable improvement, if all female schools could be furnished with suitable grounds, and instruments, for the cultivation of fruits and flowers, and every inducement offered, to engage the young ladies in this pursuit. No father, who wishes to have his daughters grow up to be healthful women, can take a surer method to secure this end. Let him set apart a portion of his yard and garden, for fruits and flowers, and see that the soil is well prepared and dug over, and all the rest may be committed to the care of the children. These would need to be provided with a light hoe and rake, a dibble, or garden trowel, a watering-pot, and means and opportunities for securing seeds, roots, buds, and grafts, all which might be done at a trifling expense. Then, with proper encouragement, and by the aid of such directions as are contained in this work, every man, who has even half an acre, could secure a small Eden around his premises.

In pursuing this amusement, children can also be led to acquire many useful habits. Early rising would, in many cases, be thus secured; and if they were required to keep their walks and borders free from weeds and rubbish, habits of order and neatness would be induced. Benevolent and social feelings could also be cultivated, by influencing children to share their fruits and flowers[Pg 252] with friends and neighbors, as well as to distribute roots and seeds to those, who have not the means of procuring them. A woman or a child, by giving seeds, or slips, or roots, to a washerwoman, or a farmer's boy, thus exciting them to love and cultivate fruits and flowers, awakens a new and refining source of enjoyment in minds, which have few resources more elevated than mere physical enjoyments. Our Saviour directs, in making feasts, to call, not the rich, who can recompense again, but the poor, who can make no returns. So children should be taught to dispense their little treasures, not alone to companions and friends, who will probably return similar favors; but to those who have no means of making any return. If the rich, who acquire a love for the enjoyments of taste, and have the means to gratify it, would aim to extend, among the poor, the cheap and simple enjoyment of fruits and flowers, our Country would soon literally "blossom as the rose."

If the ladies of a neighborhood would unite small contributions, and send a list of flower-seeds and roots to some respectable and honest florist, who would not be likely to turn them off with trash, they could divide these among themselves, so as to secure an abundant variety, at a very small expense. A bag of flower-seeds, which can be obtained, at wholesale, for four cents, would abundantly supply a whole neighborhood; and, by the gathering of seeds, in the Autumn, could be perpetuated.

Another very elevating and delightful recreation, for the young, is found in *music*. Here, the writer would protest against the common practice, in many families, of having the daughters learn to play on the piano, whether they have a taste and an ear for music, or not. A young lady, who cannot sing, and has no great fondness for music, does nothing but waste time, money, and patience, in learning to play on the piano. But all children can be taught to sing, in early childhood, if the scientific mode of teaching music, in schools, could[Pg 253] be introduced, as it is in Prussia, Germany, and Switzerland. Then, young children could read and sing music, as easily as they can read language; and might take any tune, dividing themselves into bands, and sing off, at sight, the endless variety of music which is prepared. And if parents of wealth would take pains to have teachers qualified for the purpose, as they may be at the Boston Academy, and other similar institutions, who should teach all the young children in the community, much would be done for the happiness and elevation of the rising generation. This is an amusement, which children relish, in the highest degree; and which they can enjoy, at home, in the fields, and in visits abroad.

Another domestic amusement, is, the collecting of shells, plants, and specimens in geology and mineralogy, for the formation of cabinets. If intelligent parents would procure the simpler works which have been prepared for the young, and study them, with their children, a *taste* for such recreations would soon be developed. The writer has seen young boys, of eight and ten years of age, gathering and cleaning shells from rivers, and collecting plants, and mineralogical specimens, with a delight, bordering on ecstasy; and there are few, if any, who, by proper influences, would not find this a source of ceaseless delight and improvement.

Another resource, for family diversion, is to be found in the various games played by children, and in which the joining of older members of the family is always a great advantage to both parties. All medical men unite, in declaring that nothing is more beneficial to health, than hearty laughter; and surely our benevolent Creator would not have provided risibles, and made it a source of health and enjoyment to use them, if it were a sin so to do. There has been a tendency to asceticism, on this subject, which needs to be removed. Such commands, as forbid *foolish* laughing and jesting, *"which are not convenient;"* and which forbid all idle words, and vain conversation, cannot apply to any thing, except what is foolish, vain, and useless. But jokes,[Pg 254] laughter, and sports, when used in such a degree as tends only to promote health, social feelings, and happiness, are neither vain, foolish, nor "not convenient." It is the excess of these things, and not the moderate use of them, which Scripture forbids. The prevailing temper of the mind, should be cheerful, yet serious; but there are times, when relaxation and laughter are proper for all. There is nothing better for this end, than that parents and older persons should join in the sports of childhood. Mature minds can always make such diversions more entertaining to children, and can exert a healthful moral influence over their minds; and, at the same time, can gain exercise and amusement for themselves. How lamentable, that so many fathers, who could be thus useful and happy with their children, throw away such opportunities, and wear out soul and body, in the pursuit of gain or fame!

Another resource for children, is in the exercise of mechanical skill. Fathers, by providing tools for their boys, and showing them how to make wheelbarrows, carts, sleds, and various other articles, contribute both to the physical, moral, and social, improvement of their children. And in regard to little daughters, much more can be done, in this way, than many would imagine. The writer, blessed with the example of a most ingenious and industrious mother, had not only learned, before the age of twelve, to make dolls, of various sorts and sizes, but to cut and fit and sew every article, that belongs to a doll's wardrobe. This, which was done for mere amusement, secured such a facility in mechanical pursuits, that, ever afterward, the cutting and fitting of any article of dress, for either sex, was accomplished with entire ease.

When a little girl first begins to sew, her mother can promise her a small bed and pillows, as soon as she has sewed a patch quilt for them; and then a bedstead, as soon as she has sewed the sheets and cases for pillows; and then a large doll to dress, as soon as she has made the under garments; and thus go on, till the[Pg 255] whole contents of the baby-house are earned by the needle and skill of its little owner. Thus, the task of learning to sew, will become a pleasure; and every new toy will be earned by useful exertion. A little girl can be taught, by the aid of patterns prepared for the purpose, to cut and fit all articles necessary for her doll. She can also be provided with a little wash-tub, and irons, to wash and iron, and thus keep in proper order a complete miniature domestic establishment.

Besides these recreations, there are the enjoyments secured in walking, riding, visiting, and many others which need not be recounted. Children, if trained to be healthful and industrious, will never fail to discover resources of amusement; while their guardians should lend their aid to guide and restrain them from excess.

There is need of a very great change of opinion and practice, in this Nation, in regard to the subject of social and domestic duties. Many sensible and conscientious men, spend all their time, abroad, in business, except, perhaps, an hour or so at night, when they are so fatigued, as to be unfitted for any social or intellectual enjoyment. And some of the most conscientious men in the Country, will add, to their professional business, public or benevolent enterprises, which demand time, effort, and money; and then excuse themselves for neglecting all care of their children, and efforts for their own intellectual improvement, or for the improvement of their families, by the plea, that they have no time for it. All this, arises from the want of correct notions of the binding obligation of our social and domestic duties. The main object of life, is not to secure the various gratifications of appetite or taste, but to *form such a character*, for ourselves and others, as will secure the greatest amount of present and future happiness. It is of far more consequence, then, that parents should be intelligent, social, affectionate, and agreeable, at home, and to their friends, than that they should earn money enough to live in a large house, and have handsome furniture. It is far[Pg 256] more needful, for children, that a father should attend to the formation of their character and habits, and aid in developing their social, intellectual, and moral nature, than it is, that he should earn money to furnish them with handsome clothes, and a variety of tempting food.

It will be wise for those parents, who find little time to attend to their children, or to seek amusement and enjoyment in the domestic and social circle, because their time is so much occupied with public cares or benevolent objects, to inquire, whether their first duty is not to train up their own families, to be useful members of society. A man, who neglects the mind and morals of his children, to take care of the public, is in great danger of coming under a similar condemnation, to that of him, who, neglecting to provide for his own household, has "denied the faith, and is worse than an infidel."

There are husbands and fathers, who conscientiously subtract time from their business, to spend at home, in reading with their wives and children, and in domestic amusements which at once refresh and improve. The children of such parents will grow up with a love of home and kindred, which will be the greatest safeguard against future temptations, as well as the purest source of earthly enjoyment.

There are families, also, who make it a definite object to keep up family attachments, after the children are scattered abroad; and, in some cases, secure the means for doing this, by saving money, which would otherwise have been spent for superfluities of food or dress. Some families have adopted, for this end, a practice, which if widely imitated, would be productive of extensive benefit. The method is this. On the first day of each month, some member of the family, at each extreme point of dispersion, takes a folio sheet, and fills a part of a page. This is sealed and mailed to the next family, who read it, add another contribution, and then mail it to the next.

Thus the family circular,[Pg 257] once a month, goes from each extreme, to all the members of a widely-dispersed family, and each member becomes a sharer in the joys, sorrows, plans, and pursuits, of all the rest. At the same time, frequent family meetings are sought; and the expense, thus incurred, is cheerfully met by retrenchments in other directions. The sacrifice of some unnecessary physical indulgence, (such, for instance, as the use of tea and coffee,) will often purchase many social and domestic enjoyments, a thousand times more elevating and delightful, than the retrenched luxury.

There is no social duty, which the Supreme Lawgiver more strenuously urges, than hospitality and kindness to strangers, who are classed with the widow and the fatherless, as the special objects of Divine tenderness. There are some reasons, why this duty peculiarly demands attention from the American people.

Reverses of fortune, in this land, are so frequent and unexpected, and the habits of the people are so migratory, that there are very many in every part of the Country, who, having seen all their temporal plans and hopes crushed, are now pining among strangers, bereft of wonted comforts, without friends, and without the sympathy and society, so needful to wounded spirits. Such, too frequently, sojourn long and lonely, with no comforter but Him who "knoweth the heart of a stranger."

Whenever, therefore, new comers enter a community, inquiry should immediately be made, whether they have friends and associates, to render sympathy and kind attentions; and, when there is any need for it, the ministries of kind neighborhood should immediately be offered. And it should be remembered, that the first days of a stranger's sojourn, are the most dreary, and that civility and kindness are doubled in value, by being offered at an early period.

In social gatherings, the claims of the stranger are too apt to be forgotten; especially, in cases where there are no peculiar attractions of personal appearance, or[Pg 258] talents, or high standing. Such a one should be treated with attention, *because he is a stranger;* and when communities learn to act more from principle, and less from selfish impulse, on this subject, the sacred claims of the stranger will be less frequently forgotten.

The most agreeable hospitality, to visiters, who become inmates of a family, is, that which puts them entirely at ease. This can never be the case, where the guest perceives that the order of family arrangements is essentially altered, and that time, comfort, and convenience are sacrificed, for his accommodation.

Offering the best to visiters, showing a polite regard to every wish expressed, and giving precedence to them, in all matters of comfort and convenience, can be easily combined with the easy freedom which makes the stranger feel at home; and this is the perfection of hospitable entertainment.

CHAPTER XXIV.
ON THE CONSTRUCTION OF HOUSES.

There is no point of domestic economy, which more seriously involves the health and daily comfort of American women, than the proper construction of houses. There are five particulars, to which attention should be given, in building a house; namely, economy of labor, economy of money, economy of health, economy of comfort, and good taste. Some particulars will here be pointed out, under each of these heads.

The first, respects *economy of labor*. In deciding upon the size and style of a house, the health and capacity of the housekeeper, and the probabilities of securing proper domestics, ought to be the very first consideration. If a man be uncertain as to his means for hiring service, or if he have a feeble wife, and be where properly-qualified domestics are scarce, it is very[Pg 259] poor economy to build a large house, or to live in a style which demands much labor. Every room in a house adds to the expense involved in finishing and furnishing it, and to the amount of labor spent in sweeping, dusting, cleaning floors, paint, and windows, and taking care of, and repairing, its furniture. Double the size of a house, and you double the labor of taking care of it, and so, *vice versa*. There is, in this Country, a very great want of calculation and economy, in this matter.

The arrangement of rooms, and the proper supply of conveniences, are other points, in which, economy of labor and comfort is often disregarded. For example, a kitchen will be in one story, a sitting-room in another, and the nursery in a third. Nothing is more injurious, to a feeble woman, than going up and down stairs; and yet, in order to gain two large parlors, to show to a few friends, or to strangers, immense sacrifices of health, comfort, and money, are made. If it be possible, the nursery, sitting-parlor, and kitchen, ought always to be on the same floor.

The position of wells and cisterns, and the modes of raising and carrying water, are other particulars, in which, economy of labor and comfort is sadly neglected. With half the expense usually devoted to a sideboard or sofa, the water used from a well or cistern can be so conducted, as that, by simply turning a cock, it will flow to the place where it is to be used.

A want of economy, in labor and in money, is often seen in the shape and arrangement of houses, and in the style of ornaments and furniture. A *perfect square*, encloses more rooms, at less expense, than any other shape; while it has less surface exposed to external cold, and can be most easily warmed and ventilated. And the farther a house is removed from this shape, the more the expense is increased. Wings and kitchens built out, beyond a house, very much increase expense, both in building and warming them.

Piazzas and porticoes are very expensive; and their[Pg 260] cost would secure far more comfort, if devoted to additional nursery or kitchen conveniences. Many kinds of porticoes cost as much as one additional room in the house. Houses can be so constructed, that one staircase will answer for both kitchen and parlour use, as may be seen in the engraving on page 269, (Fig. 27.) This saves the expense and labor usually devoted to a large hall and front staircase.

Much money is often worse than wasted, by finical ornaments, which are fast going out of fashion. One of the largest, most beautiful, and agreeable, houses, the writer was ever in, was finished with doors, windows, and fireplaces, in even a plainer style than any given in the subsequent drawings.

The position of fireplaces has much to do with economy of expense in warming a house. Where the fireplace is in an outer wall, one third of the heat passes out of doors, which would be retained in the house, if the chimney were within the rooms. A house, contrived like the one represented in the engraving on page 272, (Fig. 32,) which can be heated by a stove or chimney at X, may be warmed with less fuel than one of any other construction.[R]

Economy of health is often disregarded, by placing wells, cisterns, and privies, so that persons, in the perspiration of labor, or the debility of disease, are obliged to go out of doors in all weathers. Figure 35, on page 276, shows the proper arrangement of such conveniences. The placing of an outside door, for common use, in a sitting-room, as is frequent at the West and South, is detrimental to health. In such cases, children, in their sports, or persons who labor, are thrown into perspiration, by exercise, the door is thrown open, a chill ensues, and fever, bowel complaints, or bilious attacks,[Pg 261] are the result. A long window, extending down to the floor, which can be used as a door, in Summer, and be tightly closed, at the bottom, in Winter, secures all the benefits, without the evils, of an outside door.

Constructing houses, without open fireplaces in chambers, or any other mode of ventilation, is another sad violation of the economy of health. Feeble constitutions in children, and ill health to domestics, are often caused by this folly.

The *economy of comfort* is often violated, by arrangements made for domestics. Many a woman has been left to endure much hard labor and perplexity, because she chose to have money spent on handsome parlors and chambers, for company, which should have been devoted to providing a comfortable kitchen and chambers for domestics. Cramping the conveniences and comfort of a family, in order to secure elegant rooms, to show to company, is a weakness and folly, which it is hoped will every year become less common.

The construction of houses with reference to *good taste*, is a desirable, though less important, item. The beauty of a house depends very much upon propriety of proportions, color, and ornament. And it is always as cheap, and generally cheaper, to build a house in agreement with the rules of good taste, than to build an awkward and ill-proportioned one.

Plans of Houses and Domestic Conveniences.

The following plans are designed chiefly for persons in moderate circumstances, and have especial reference to young housekeepers.

Every year, as the prosperity of this Nation increases, good domestics will decrease, and young mothers are hereafter to be called to superintend and perform all branches of domestic business, to nurse children, direct ignorant domestics, attend the sick, entertain company, and fulfil all other family duties; and this, too, in a majority of cases, with delicate constitutions, or impaired health. Every man, therefore, in forming plans for a[Pg 262] future residence, and every woman who has any influence in deciding such matters, ought to make these probabilities the chief basis of their calculations.[S]

Fig.
17.

Fig. 18.
Ground-plan.

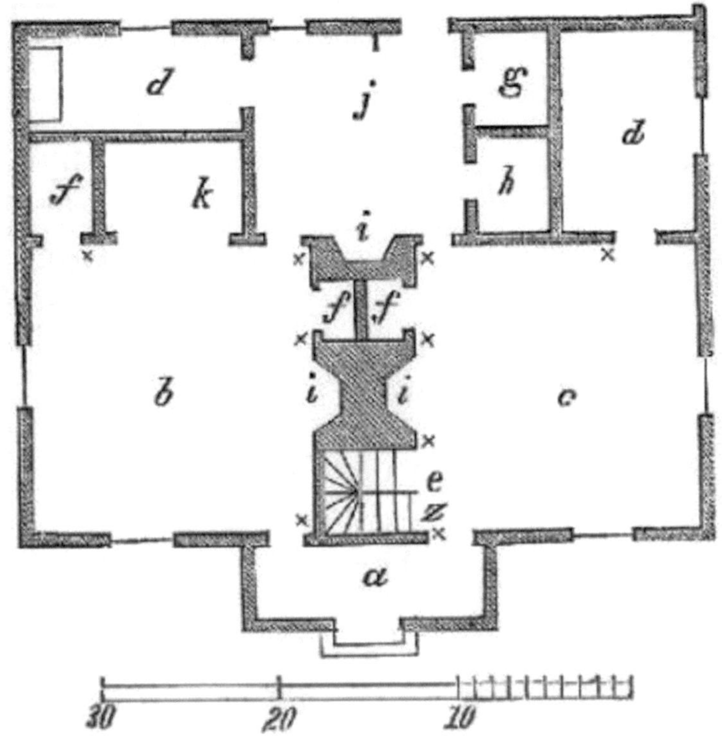

Scale of Feet.

- a, Porch.
- b, Parlor, 15 by 16 feet.
- c, Dining-room, 15 by 16 feet.
- d, d, Small Bedrooms.
- e, Stairs.
- f, f, f, Closets.
- g, Pantry.
- h, Store-closet.
- i, i, i, Fireplaces.
- j, Kitchen.
- k, Bedpress.
- z, Cellar door.

The[Pg 263] plan, exhibited in Figures 17, and 18, is that of a cottage, whose chief exterior beauty is its fine proportions. It should be painted white.

Fig. 17, is the *elevation*, or the front view of the exterior. Fig. 18, is the ground-plan, in which, an entire break in the wall, represents a door, and a break with a line across it, a window. When a cross x is put by a door, it indicates into which room the door swings, and where the hinges should be put, as the comfort of a fireside very much depends on the way in which the doors are hung. A scale of measurement is given at the bottom of the drawings, by which, the size of all parts can be measured. The ten small divisions, are each one foot. The longest divisions are ten feet each.

In the ground-plan, (Fig. 18,) *a*, is the porch, which projects enough to afford an entrance to the two adjacent rooms, and thus avoids the evil of an outside door to a sitting-room. If a door be wanted in these rooms, the front windows can be made to extend down to the floor, so as to serve as doors in Summer, and be tightly closed in Winter. The parlor, *b*, has the bedpress, *k*, and the closet, *f*, adjoining it. Figure 19 is intended to represent this side of the room.

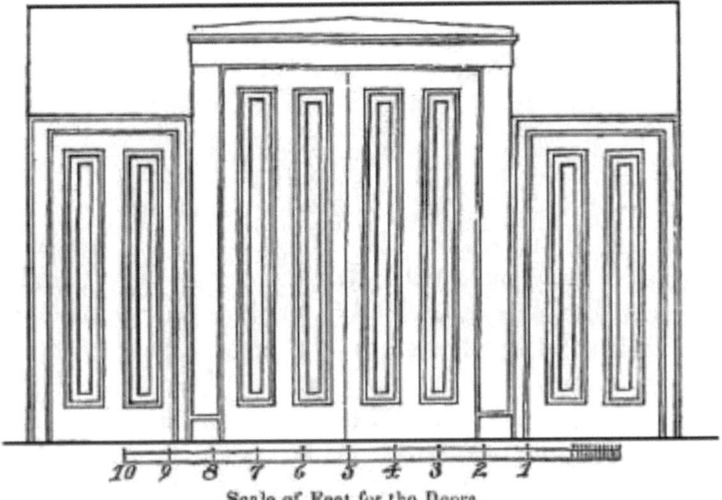

[Pg 264]

The two large doors, in the centre, open into the bedpress, and one of the smaller ones into the closet, *f*. The other, can either be a false door, in order to secure symmetry, or else a real one, opening into the kitchen, *j*.

A room, thus arranged, can be made to serve as a genteel parlor, for company, during the day, when all these doors can be closed. At night, the doors of the bedpress being opened, it is changed to an airy bedroom, while the closets, *f, f,* serve to conceal all accommodations pertaining to a bedroom. The bedpress is just large enough to receive a bed; and under it, if need be, might be placed a trucklebed, for young children. The eating-room, *c*, has the small bedroom, *d*, adjoining it, which, by leaving the door open, at night, will be sufficiently airy for a sleeping-room. The kitchen, *j*, has a smaller bedroom, *d*, attached to it, which will hold a narrow single bed for a domestic; and, if need be, a narrow trucklebed under it, for a child. The staircase to the garret, can either be placed in the eating-room, or in the small entry. A plan for back accommodations is shown in

Fig. 35, (page 276.) These should be placed in the rear of the kitchen, so as not to cover the window.

A house like this, will conveniently accommodate a family of six or eight persons; but some economy and contrivance will be needed, in storing away articles of dress and bedclothing. For this end, in the bedpress, *k,* of the parlor, *b,* (Fig. 18,) a wide shelf may be placed, two feet from the ceiling, where winter bedding, or folded clothing, can be stowed, while a short curtain in front, hung from the wall, will give a tidy look, and keep out dust. Under this shelf, if need be, pegs can be placed, to hold other articles; and a curtain be hung from the edge of the shelf, to conceal and protect them. Both the closets, *f, f,* should have shelves and drawers. The garret can have a window inserted in the roof, and thus be made serviceable for storage.[Pg 265]

Fig.
20.

Figure 20 represents a fireplace and mantelpiece, in a style corresponding with the doors.

Such a cottage as this, could be built for from five hundred to nine hundred dollars, according as the expense of labor in the place, and the excellence of the materials and labor, may vary.

Figures 21 and 22, show the elevation and ground-plan of a cottage, in which the rooms are rather more agreeably arranged, than in the former plan. The elevation, (Fig. 21,) has a piazza, running across the whole front. This would cost nearly two hundred dollars; and, for this sum, another story might be added. An architect told the writer, that he could build[Pg 266] the two-story house, (Fig. 23 and 24,) without a piazza, for the same sum, as this cottage, *with* one. This shows the poor economy of these appendages.

The ground-plan, (Fig. 22,) will be understood, from the explanation appended to it.

Fig.
22.

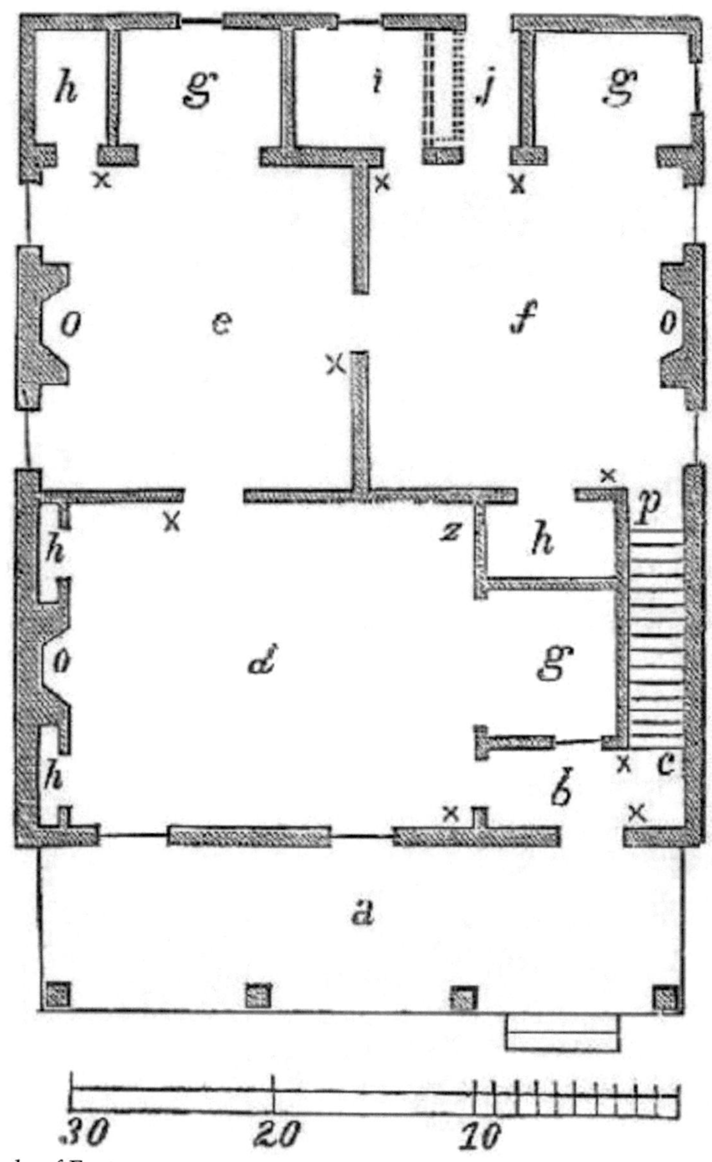

Scale of Feet.

.30   20   10

274

- a, Porch.
- b, Entry.
- c, Stairs.
- d, Parlor, 16 by 20 feet.
- e, Dining-room, 16 by 16 feet.
- f, Kitchen.

- g, g, g, Bedpresses.
- h, h, h, h, Closets.
- i, Store-closet.
- j, Back entry and Sink.
- p, Cellar stairs.
- o, o, o, Fireplaces.

The parlor, *d*, is designed to have the doors (shown in Fig. 19) placed at the end, where is the bedpress, *g*. This will make it a handsome parlor, by day, and yet[Pg 267] allow it to be used as a bedroom, at night. The bedpresses, in the other rooms, can have less expensive doors. A window is put in each bedpress, to secure proper ventilation. These should be opened, to air the bed, on leaving it. These can be fitted up with shelves, pegs, and curtains, as before described. If the elevation of the first cottage be preferred to this, as being less expensive, it can be used, by altering it a little; thus, instead of the projection for the entry, make a slight projection, of the width of one brick, to preserve the same general outside appearance. Let the windows extend down to the floor, and the beauty of symmetry will also be preserved.

Fig. 23.
Ground-plan.

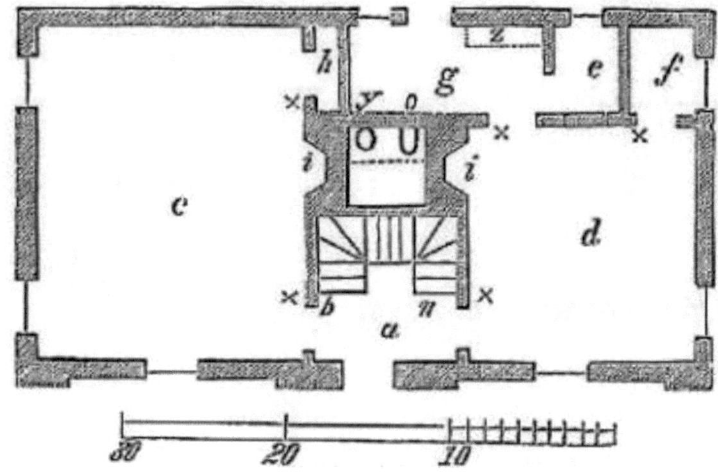

Scale of Feet.

- a, Entry.
- b, Stairs.
- c, Parlor, 16 by 20 feet.
- d, Kitchen, 14 by 14 feet.
- e, Store-closet.
- f, Pantry.
- g, Sinkroom.
- h, Closet.
- i, i, Fireplaces.
- n, Cellar door.
- o, Oven.
- y, Furnace.
- z, Sink.

Fig. 24.
Second Story.

276

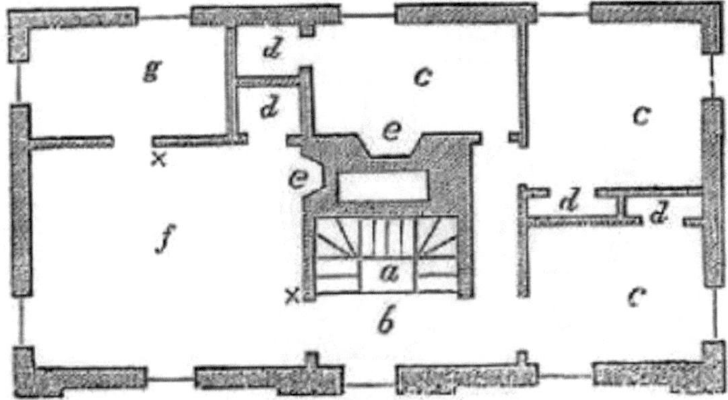

- a, Stairs.
- b, Passage.
- c, c, c, Bedrooms.
- d, d, d, d, Closets.
- e, e, Fireplaces.
- f, Nursery.
- g, Room for young children.

[Pg 268] The plans, shown in Fig. 23 and 24, are designed for families, where most domestic labor is to be done without the aid of domestics. The parlor, *c*, is for a sitting-room, and for company. The room, *d*, is the eating-room; where, also, the ironing and other nicer family work can be done. In the small room, *g*, either an oven and boiler, or a cooking-stove, can be placed. The elevation, shown in Fig. 25, is designed for the front of this house.

Fig.
25.

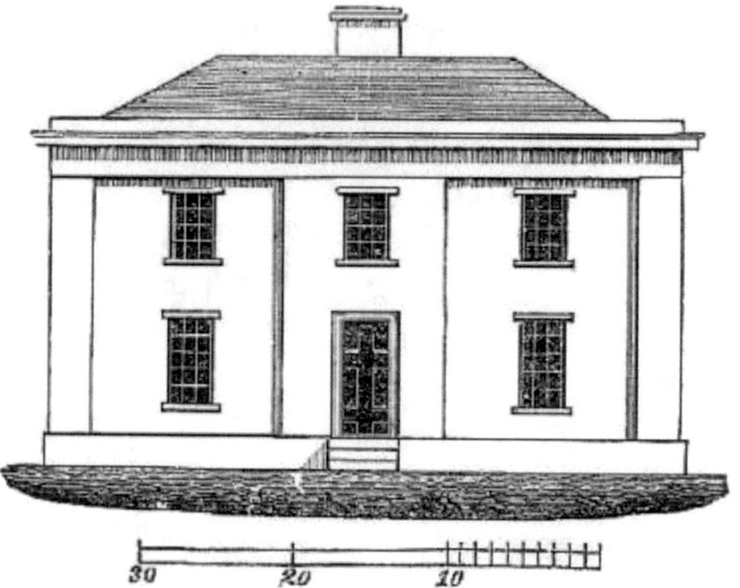

30
20
10

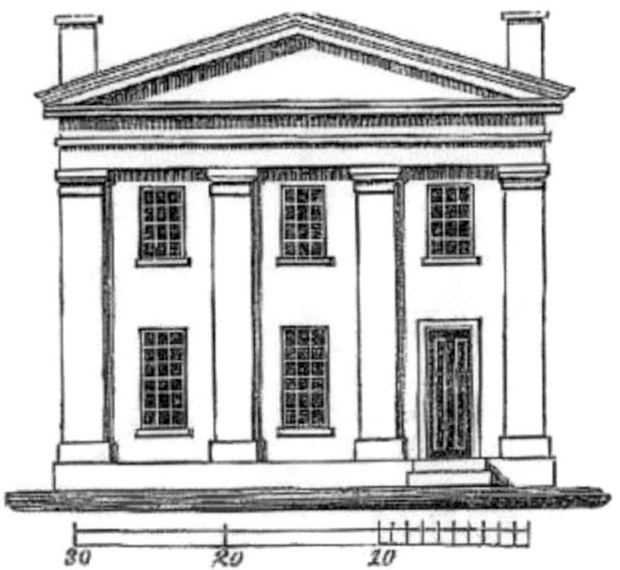

[Pg 269]Figures 27 and 28, are plans of a two-story house, on a larger scale, with a concealed staircase, for front and back use. The elevation, Fig. 26, is designed for this plan.

Fig. 27.
Ground-plan.

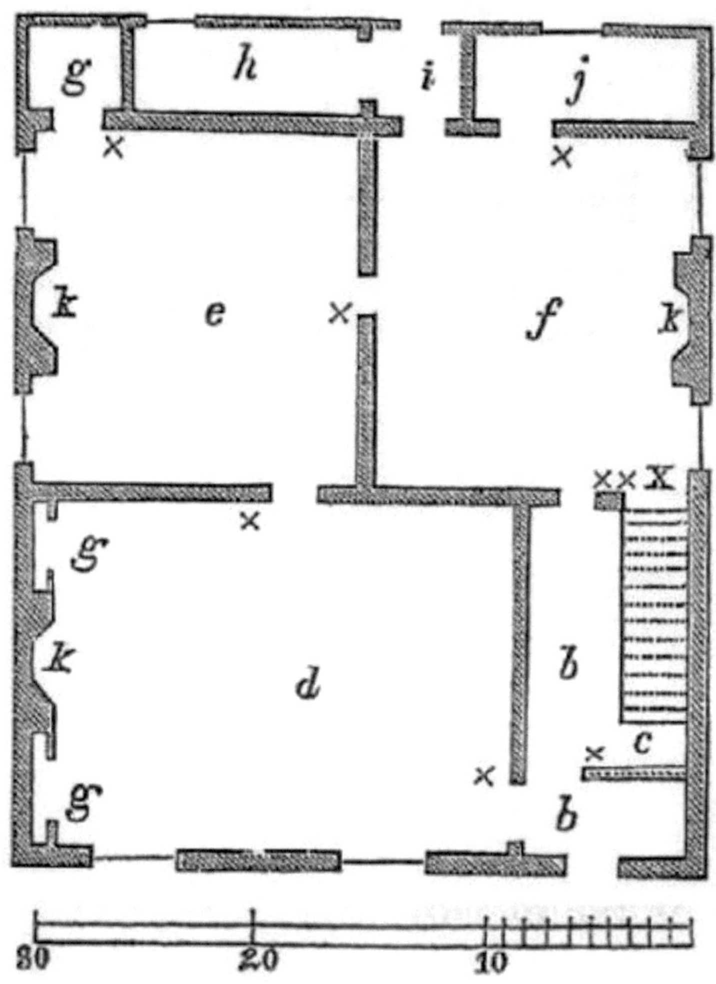

- b, b, Entry.
- c, Stairs.
- d, Parlor, 16 by 20 feet.
- e, Dining-room, 15 by 16 feet.
- f, Kitchen, 15 by 16 feet.

- g, g, g, Closets.
- h, Store-closet.
- i, Back entry.
- j, Pantry.
- k, k, k, Fireplaces.
- x, Cellar stairs.

Fig. 28.
Second Story.

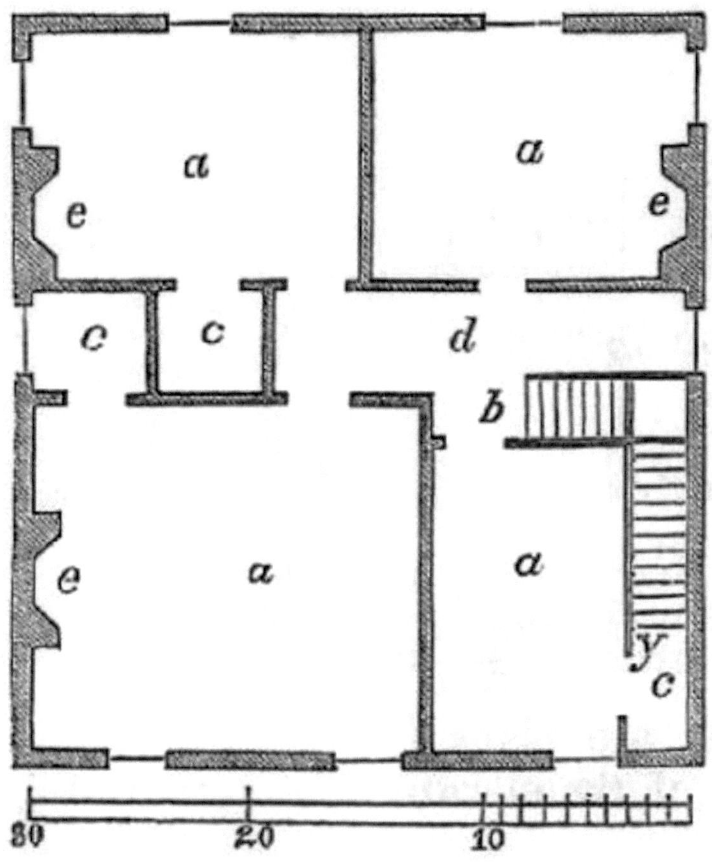

- a, a, a, a, Bedrooms.
- b, Stairs.
- c, c, c, Closets.
- d, Passage.
- e, e, e, Fireplaces.
- y, Garret stairs.

[Pg 270]

Fig. 29.
Ground-floor.

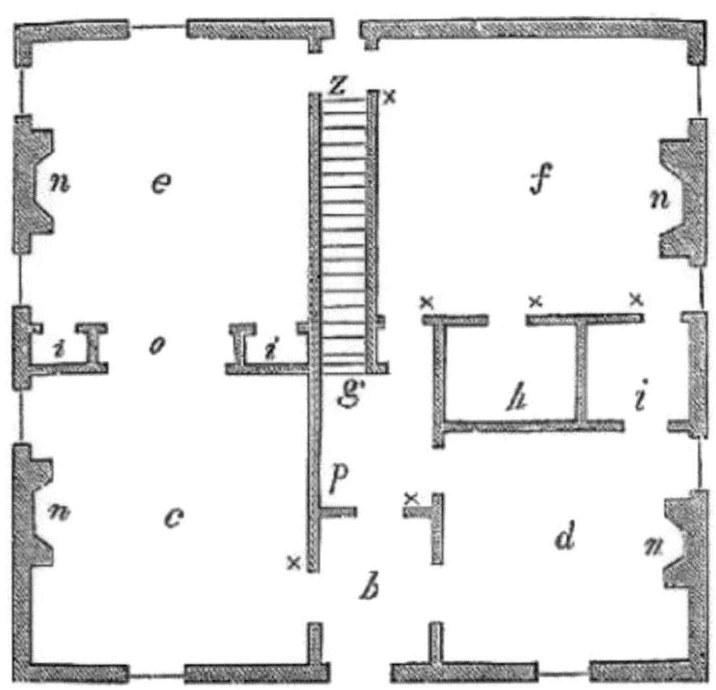

- b, Entry.
- c, Parlor, 17 by 17 feet.
- d, Dining-room, 13 by 15 feet.
- e, Parlor or Bedroom,
  17 by 17 feet.
- f, Kitchen, 19 by 17 feet.
- g, Stairs.
- h, Store-closet.
- i, i, i, Closets.

- n, n, n, n, Fireplaces.
- o, Folding-doors.
- p, Pegs for over-garments.
- z, Cellar stairs.

Fig. 30.
Second Story.

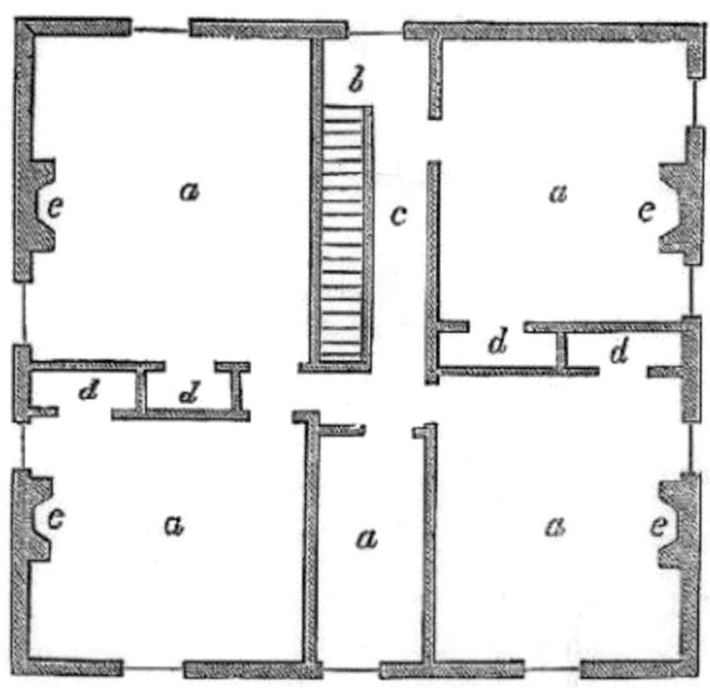

- a, a, a, a, a, Bedrooms.
- b, Stairs.
- c, Passage.
- d, d, d, d, Closets.
- e, e, e, e, Fireplaces.

Figures 29 and 30, are plans for a larger house, which can have either of the elevations, Fig. 25 or 26,[Pg 271] adapted to it. These also have a concealed staircase, for front and back use. If a nursery, or bedroom, is wished, on the ground-floor, the back parlor, *e*, can be taken; in which case, the closets, *i*, *i*, are very useful. To prevent noise from reaching the front parlor, two sets of folding-doors, each side of the passage, *o*, could be placed. With this arrangement, these rooms could be used, sometimes as two parlors, opening into each other, by folding doors, and at other times, as a nursery and parlor. In this plan, the storeroom, *h*, and china-closet, *i*, between the kitchen and eating-room, are a great convenience.

Figures 31 and 32, present the plan of a Gothic cottage, which secures the most economy of *labor* and *expense*, with the greatest amount of *convenience and comfort*, which the writer has ever seen.

Fig.
31.

Scale of Feet.

The elevation, (Fig. 31,) exhibits the front view. It has a recess in the central part, under which, is the door, with a window on each side of it. This forms a piazza; and into this, and a similar one at the back of the house, the two centre parlors open.[Pg 272]

Fig. 32.

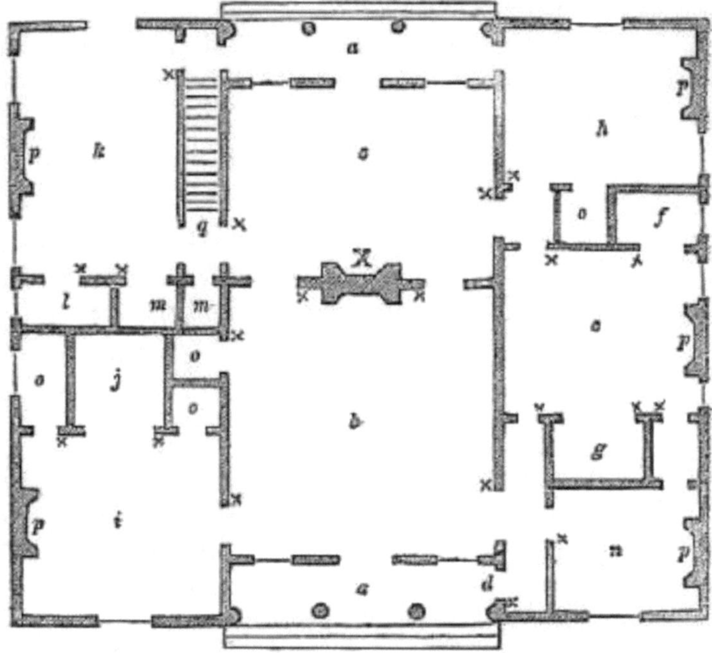

In the centre of the house, (see Fig. 32,) are the two parlors, *b* and *c*; the back one to be used as an eating-room. At X, can be placed, either a chimney, with doors on each side of the fireplace, or, (which is the most agreeable,) folding-doors, which can be thrown open in Summer, thus making a large saloon, through the house, from one piazza to the other. In this case, the parlors are warmed by a large stove, set near the folding-doors, which would easily warm both parlors and one or two adjacent rooms. In Winter, the outside doors, opening to the piazzas, should be fastened and calked, and the side entry, at *d*, be used. At *e*, is the nursery, with the bedpress, *g*, which, being closed by day, makes a retired parlor for the mother. At *n*, is the children's playroom and sleeping-room, adjoining the

286

mother's room. At *k*, is the kitchen, adjacent to the eating-room, with the storeroom, *e*, and the closets, *m, m*, one for the eating-room, and one for the kitchen utensils. At *i*, is a parlor, which can be used for a study or library, by the master of the family; while the[Pg 273] adjacent bedpress, *j*, renders it a convenient lodging-room, for guests. Another lodging-room, is at *h*; and in the attic, is space enough for several comfortable lodging-rooms. A window in the roof, on the front and back, like the one on Wadsworth's Cottage, (Fig. 33,) could be placed over the front door, to light the chambers in the attic. A double roof in the attic, with a current of air between, secures cool chambers. The closets are marked *o*, and the fireplaces *p*. The stairs to the attic are at *q*. By this arrangement, the house-keeper has her parlor, sleeping-room, nursery, and kitchen, on the same floor, while the rooms with bedpresses, enable her to increase either parlors or lodging-rooms, at pleasure, without involving the care of a very large and expensive house.

Figure 33, is the representation of a cottage, built by Daniel Wadsworth, Esq., in the vicinity of Hartford, Connecticut; and is on a plan, which, though much smaller, is very similar to the plan represented in Fig. 32. It serves to show the manner in which the *roofs* should be arranged, in Fig. 31, which, being seen exactly in front, does not give any idea of the mode of this arrangement. The elevation of Wadsworth's cottage, could be taken for the ground-plan shown in Fig. 32, if it be preferred to the other.

Both this cottage, and all the other plans, require a woodhouse, and the conveniences connected with it, which are represented in Fig. 35, (page 276.) For these Gothic cottages, an appendage of this sort should be in keeping with the rest, having windows, like those in the little Summer-house in the drawing, and battlements, as on the top of the wings of the barn. The ornaments on the front of the cottage, and the pillars of the portico, made simply of the trunks of small trees, give a beautiful rural finish, and their expense is trifling. In this picture, the trees could not be placed as they are in reality, because they would hide the buildings.[Pg 274]

[Pg 275]In arranging yards and grounds, the house should be set back, as in the drawing of Wadsworth's cottage; and, instead of planting shade-trees in straight lines, or scattering them about, as single trees, they should be arranged in clusters, with large openings for turf, flowers, and shrubbery, which never flourish well under the shade and dropping of trees. This also secures spots of dark and cool shade, even when trees are young.

In arranging shade-trees tastefully around such a place, a large cluster might be placed on each side of the gate; another on the circular grass-plot, at the side of the house; another at a front corner; and another at a back corner. Shrubbery, along the walks, and on the circular plot, in front, and flowers close to the house, would look well. The barn, also, should have clusters of trees near it; and occasional single trees, on the lawn, would give the graceful ease and variety seen in nature.

Figure 34, represents the accommodations for securing water with the least labor. It is designed for a well or cistern under ground. The reservoir, R, may be a half hogshead, or something larger, which may be filled once a day, from the pump, by a man, or boy.

Fig.
34.

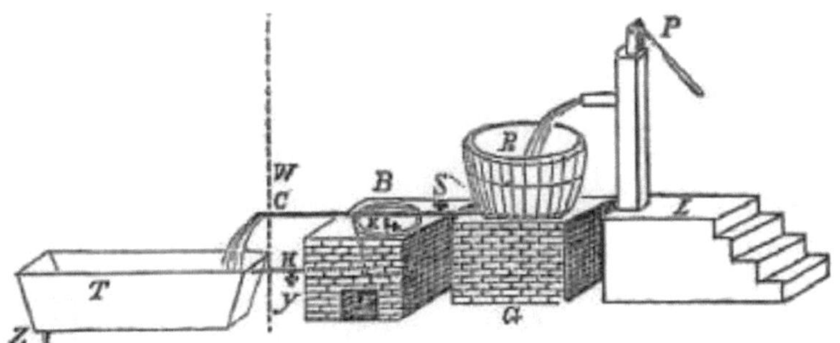

P, Pump. L, Steps to use when pumping. R, Reservoir. G, Brick-work to raise the Reservoir. B, A large Boiler. F, Furnace, beneath the Boiler. C, Conductor of cold water. H, Conductor of hot water. K, Cock for letting cold water into the Boiler. S, Pipe to conduct cold water to a cock over the kitchen sink. T, Bathing-tub, which receives cold water from the Conductor, C, and hot water from the Conductor, H. W, Partition separating the Bathing-room from the Wash-room. Y, Cock to draw off hot water. Z, Plug to let off the water from the Bathing-tub into a drain.

[Pg 276]The conductor, C, should be a lead pipe, which, instead of going over the boiler, should be bent along behind it. From S, a branch sets off, which conducts the cold water to the sink in the kitchen, where it discharges with a cock. H, is a conductor from the lower part of the boiler, made of copper, or some metal not melted by great heat; and at Y, a cock is placed, to draw off hot water. Then the conductor passes to the bathing-tub, where is another cock. At Z, the water is let off from the bathing-tub. By this arrangement, great quantities of hot and cold water can be used, with no labor in carrying, and with very little labor in raising it.

In case a cistern is built above ground, it can be placed as the reservoir is, and then all the labor of pumping is saved.

Fig. 35.

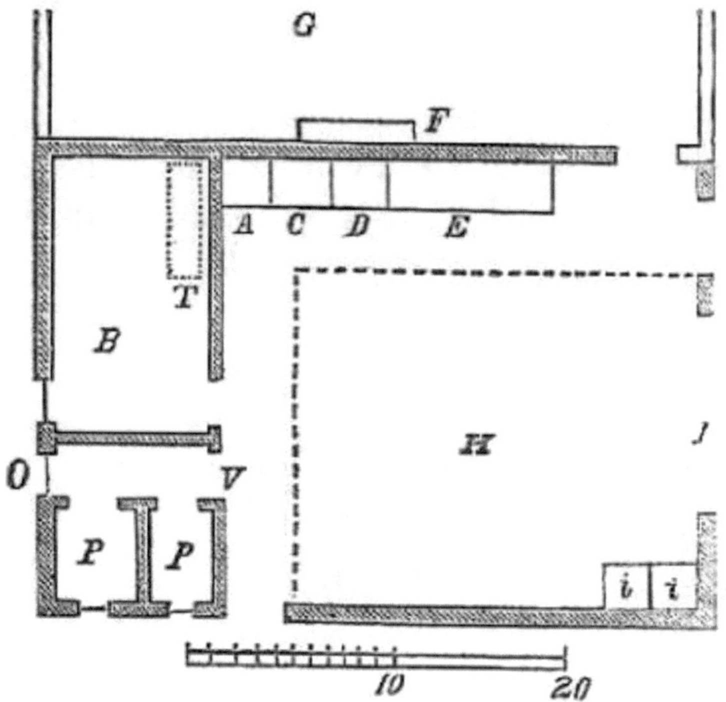

Scale of Feet.

- A, Boiler and furnace.
- B, Bathing-room.
- C, Reservoir.
- D, Pump.
- E, Wash-form.
- F, Sink.
- G, Kitchen.
- H, Woodpile.
- I, Large doors.
- i,i, Bins for coal and ashes.
- O, Window.
- P, P, Privies.

- T, Bathing-tub.
- V, Door.

Fig. 35, is the plan of a building for back-door accommodations. At *A*, *C*, *D*, *E*, are accommodations shown in Fig. 34. The bathing-room is adjacent to the boiler and reservoir, to receive the water. The privy, *P*, *P*, should have two apartments, as indispensable [Pg 277]to healthful habits in a family. A window should be placed at *O*, and a door, with springs or a weight to keep it shut, should be at *V*. Keeping the window open, and the door shut, will prevent any disagreeable effects in the house. At *G*, is the kitchen, and at *F*, the sink, which should have a conductor and cock from the reservoir. *H*, is the place for wood, where it should in Summer be stored for Winter. A bin, for coal, and also a brick receiver, for ashes, should be in this part. Every woman should use her influence to secure all these conveniences; even if it involves the sacrifice of the piazza, or "the best parlor."

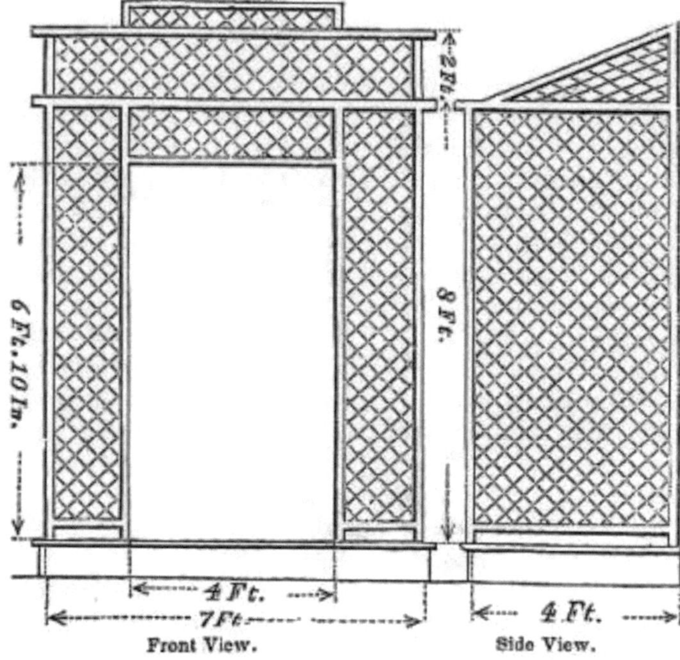

Fig. 36, is a latticed portico, which is cheap, and answers all the purposes of a more expensive one. It should be solid, overhead, to turn off the rain, and creepers should be trained over it. A simple latticed[Pg 278] arch, over a door, covered with creepers, is very cheap, and serves instead of an expensive portico.

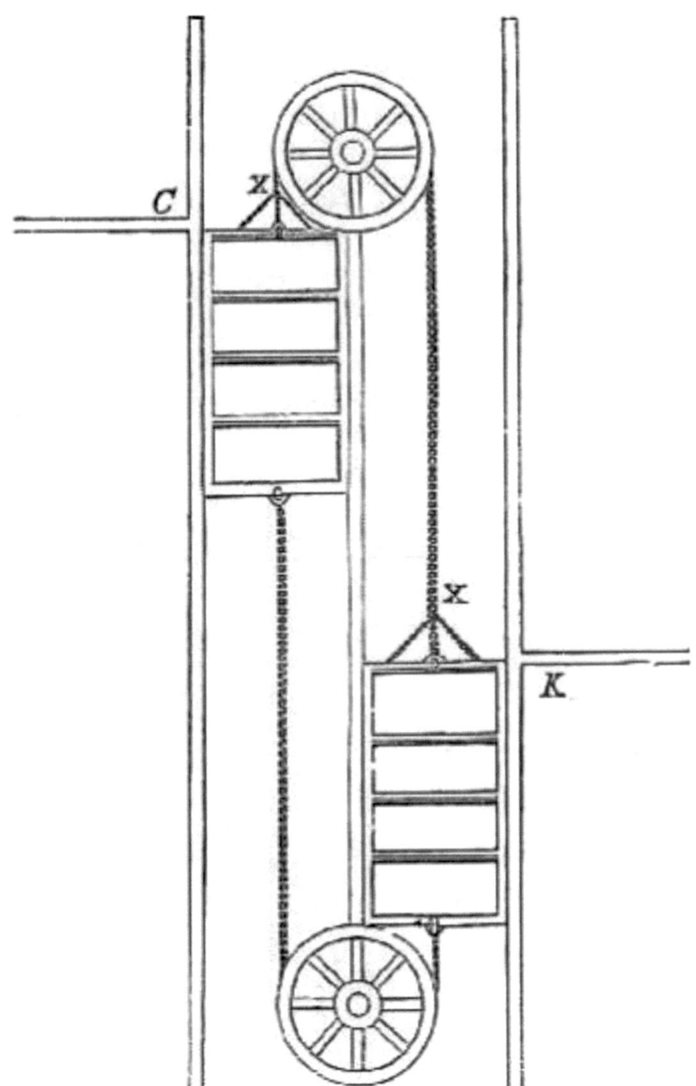

C, Parlor ceiling. K, Kitchen ceiling.

Fig. 37, represents a *sliding closet*, or *dumb waiter*, a convenience which saves much labor, when the kitchen is in the basement. The two closets should be made wide, and broad enough to receive a common waiter. The chain, or rope, which passes over the wheels, should branch, at X, so as to keep the closet from rubbing in its movements, when the dishes are[Pg 279] not set exactly in the middle, or are of unequal weights. By this method, almost every thing needed to pass between the kitchen and parlor can be sent up and down, without any steps. If the kitchen is not directly under the eating-room, the sliding closet can be placed in the vicinity of one or both. Where the place is not wide enough for two closets like these, they can be made wider than they are long, say one foot and six inches long, and three feet wide. A strip of wood, an inch broad, should be fastened on the front and back of the shelves, to prevent the dishes from being broken when they are set on carelessly.

There is nothing, which so much improves the appearance of a house and the premises, as painting or whitewashing the tenements and fences. The following receipts for whitewashing, answer the same purpose for wood, brick, and stone, as oil-paint, and are much cheaper. The first, is the receipt used for the President's house, at Washington, improved by further experiments. The second, is a cheaper one, which the writer has known to succeed, in a variety of cases, lasting as long, and looking as well, as white oil-paint.

Receipt.

Take half a bushel of unslacked lime, and slack it with boiling water, covering it, during the process. Strain it, and add a peck of salt, dissolved in warm water; three pounds of ground rice, boiled to a thin paste, put in boiling hot; half a pound of powdered Spanish whiting; and a pound of clear glue, dissolved in warm water. Mix, and let it stand several days. Heat it in a kettle, on a portable furnace, and apply it as hot as possible, with a painter's or whitewash-brush.

Another.

Make whitewash, in the usual way, except that the water used should be hot, and nearly saturated with salt. Then stir in four handfuls of fine sand, to make[Pg 280] it thick like cream. Coloring matter can be added to both, making a light stone-color, a cream-color, or a light buff, which are most suitable for buildings.

FOOTNOTES:

[R] Many houses are now heated, by a furnace in the cellar, which receives pure air from out of doors, heats it, and sends it into several rooms, while water is evaporated to prevent the air from becoming dry. The most perfect one the writer has seen, is constructed by Mr. Fowler, of Hartford. This method secures well-ventilated rooms, and is very economical, where several rooms are to be warmed.

[S] Those, who are amateurs in architecture, in judging of these designs, must take into consideration, that this is a work on domestic *economy*, and that matters of taste, have necessarily been made subordinate to points, involving economy of health, comfort, and expense. Still, it is believed, that good taste has been essentially preserved, in most of these designs.

CHAPTER XXV.
ON FIRES AND LIGHTS.

A shallow fireplace saves wood, and gives out more heat than a deeper one. A false back, of brick, may be put up in a deep fireplace. Hooks, for holding up the shovel and tongs, a hearth-brush and bellows, and brass knobs to hang them on, should be furnished to every fireplace. An iron bar, across the andirons, aids in keeping the fire safe, and in good order. Steel furniture is more genteel, and more easily kept in order, than that made of brass.

Use green wood, for logs, and mix green and dry wood for the fire; and then the woodpile will last much longer. Walnut, maple, hickory, and oak, wood, are best, chestnut or hemlock is bad, because it snaps. Do not buy a load, in which there are many crooked sticks. Learn how to measure and calculate the solid contents of a load, so as not to be cheated. Have all your wood split, and piled under cover, for Winter. Have the green wood logs in one pile, dry wood in another, oven-wood in another, kindlings and chips in another, and a supply of charcoal to use for broiling and ironing, in another place. Have a brick bin, for ashes, and never allow them to be put in wood. When quitting fires, at night, never leave a burning stick across the andirons, nor on its end, without quenching it. See that no fire adheres to the broom or brush; remove all articles from the fire, and have two pails, filled with water, in the kitchen, where they will not freeze.

[Pg 281]

Stoves and Grates.

Rooms, heated by stoves, should always have some opening for the admission of fresh air, or they will be injurious to health. The dryness of the air, which they occasion, should be remedied, either by placing a vessel, filled with water, on the stove, or by hooking a long and narrow pan, filled with water, in front of the grate; otherwise, the lungs or eyes may be injured. A large number of plants in a room, prevents this dryness of the air. Openings for pipes, through floors, partitions, or fireboards, should be surrounded by tin, to prevent their taking fire. Lengthening a pipe, will increase its draught.

For those, who use *anthracite coal,* that which is broken or screened, is best for grates, and the nut-coal, for small stoves. Three tons are sufficient, in the Middle States, and four tons in the Northern, to keep one fire through the Winter. That which is bright, hard, and clean, is best; and that which is soft, porous, and covered with damp dust, is poor. It will be well to provide two barrels of charcoal, for kindling, to every ton of anthracite coal. Grates, for *bituminous* coal, should have a flue nearly as deep as the grate; and the bars should be round, and not close together. The better draught there is, the less coal-dust is made. Every grate should be furnished with a poker, shovel, tongs, blower, coal-scuttle, and holder for the blower. The latter may be made of woollen, covered with old silk, and hung near the fire.

Coal-stoves should be carefully put up, as cracks, in the pipe, especially in sleeping rooms, are dangerous.

On Lights.

Lamps are better than candles, as they give a steadier light, and do not scatter grease, like tallow candles. The best oil, is clear, and nearly colorless. Winter-strained oil should be used in cold weather. Lard is a good substitute for oil, for astral and other large lamps.[Pg 282] It is cheaper, burns clearer, and has a less disagreeable smell. It will not burn so well in small lamps, as in large ones. Melt it every morning, in an old pitcher, kept for the purpose. Oil, long kept, grows thick, and does not burn well. It is therefore best not to buy it in large quantities. It should never be left standing in lamps, for several days, as this spoils it, and often injures the lamps. Camphine is a kind of oil manufactured in New York, which does not smell disagreeably, nor make grease-spots, and gives a brighter light than the best oil. Cleanse the insides of lamps and oil-cans, with pearlash-water. Be careful to drain them well, and not to let any gilding, or bronze, be injured by the pearlash-water coming in contact with it. Put one tablespoonful of pearlash to one quart of water.

The care of lamps requires so much attention and discretion, that many ladies choose to do this work, themselves, rather than trust it with domestics. To do it properly, provide the following things: — An old waiter, to hold all the articles used; a lamp-filler, with a spout, small at the end, and turned up to prevent oil from dripping; a ball of wickyarn, and a basket to hold it; a lamp-trimmer, made for the purpose, or a pair of *sharp* scissors; a small soap-cup and soap; some pearlash, in a broad-mouthed bottle; and several soft cloths, to wash the articles, and towels, to wipe them. If every thing, after being used, is cleansed from oil, and then kept neatly, it will not be so unpleasant a task, as it usually is, to take care of lamps.

Wash the shade of an astral lamp, once a week, and the glass chimney oftener. Take the lamp to pieces, and cleanse it, once a month. Keep dry fingers, in trimming lamps. To raise the wick of an astral lamp, turn it to the right; to lower it, turn it to the left. Trim it, after it has been once used; and, in lighting it, raise it to the proper height, as soon as may be, or it will either smoke, or form a crust. Renew the wick, when only an inch and a half long. Close-woven wicks are better than those which are loose. Dipping[Pg 283] wicks in vinegar, makes them burn clearer than they otherwise would. Plain shades do not injure the eyes, like cut ones; and prints and pictures appear better by them, than by the others. Lamps should be lighted with a strip of folded or rolled paper, kept on the mantel-piece. Weak eyes should always be shaded from the lights. Small screens, made for the purpose, should be kept at hand. A person with weak eyes, can use them, safely, much longer, when they are shaded from the glare of the light, than if they are not so. Fill the entry-lamp, every day, and cleanse and fill night-lanterns, twice a week, if used often. Provide small, one-wicked lamps, to carry about; and broad-bottomed lamps, for the kitchen, as these are not easily upset.

A good night-lamp is made, with a small one-wicked lamp and a roll of tin to set over it. Have some holes made in the bottom of this cover, and it can then be used to heat articles. Very cheap floating tapers, can be bought, to burn in a teacup of oil through the night.

Wickyarn, drawn repeatedly through melted wax, till stiff and smooth, makes a good taper, for use in sealing letters. It can be twined in fanciful forms, and kept on the writing-table.

To make Candles.

The nicest candles, are run in moulds. For this purpose, melt together one quarter of a pound of white wax, one quarter of an ounce of camphor, two ounces of alum, and ten ounces of suet or mutton tallow. Soak the wicks, in lime-water and saltpetre, and, when dry, fix them in the moulds, and pour in the melted tallow. Let them remain one night, to cool, then warm them, a little, to loosen them, draw them out, and, when hard, put them in a box, in a dry and cool place.

To make dipped candles, cut the wicks of the right length, double them over rods, and twist them. They should first be dipped in lime-water, or vinegar, and dried. Melt the tallow in a large kettle, filling it[Pg 284] to the top with hot water, when the tallow is melted. Put in wax, and powdered alum, to harden them. Keep the tallow hot, over a portable furnace, and fill up the kettle, with hot water, as fast as the tallow is used up. Lay two long strips of narrow board, on which to hang the rods; and set flat pans under, on the floor, to catch the grease. Take several rods at once, and wet the wicks in the tallow; and, when cool, straighten and smooth them. Then dip them, as fast as they cool, until they become of the proper size. Plunge them obliquely, and not perpendicularly; and when the bottoms are too large, hold them in the hot grease, till a part melts off. Let them remain one night, to cool; then cut off the bottoms, and keep them in a dry, cool place. Cheap lights are made, by dipping rushes in tallow.

CHAPTER XXVI.
ON WASHING.

There is nothing, which tends more effectually to secure good washing, than a full supply of all conveniences; and among these, none is more important, than an abundance of warm and cold water: but, if this be obtained, and heated, at a great expense of time and labor, it will be used in stinted measure. The accommodations described on page 275, (Fig. 34,) are very convenient in this respect.

Articles to be provided for Washing.

A plenty of soft water is a very important item. When this cannot be had, ley or soda can be put in hard water, to soften it; care being used not to put in so much, as to injure the hands and clothes. Two wash-forms are needed; one for the two tubs in which to put the suds, and the other for blueing and starching-tubs. Four tubs, of different sizes, are necessary; also,[Pg 285] a large *wooden* dipper, (as metal is apt to rust;) two or three pails; a grooved wash-board; a clothes-line, (sea-grass, or horse-hair is best;) a wash-stick to move clothes, when boiling, and a wooden fork to take them out. Soap-dishes, made to hook on the tubs, save soap and time. Provide, also, a clothes-bag, in which to boil clothes; an indigo-bag, of double flannel; a starch-strainer, of coarse linen; a bottle of ox-gall for cali-coes; a supply of starch, neither sour nor musty; several dozens of clothes-pins, which are cleft sticks, used to fasten clothes on the line; a bottle of dissolved gum Arabic; two clothes-baskets; and a brass or copper kettle, for boiling clothes, as iron is apt to rust. A closet, for keeping all these things, is a great convenience. It may be made six feet high, three feet deep, and four feet wide. The tubs and pails can be set on the bottom of this, on their sides, one within another. Four feet from the bottom, have a shelf placed, on which to put the basket of clothes-pins, the line, soap-dishes, dipper, and clothes-fork. Above this, have another shelf, for the bottles, boxes, &c. The shelves should reach out only half way from the back, and nails should be put at the sides, for hanging the wash-stick, clothes-bag, starch-bag, and indigo-bag. The ironing-conveniences might be kept in the same closet, by having the lower shelf raised a little, and putting a deep drawer under it, to hold the ironing-sheets, holders, &c. A lock and key should be put on the closet. If the mistress of the family requests the washerwoman to notify her, when she is through, and then ascertains if all these articles are put in their places, it will prove useful. Tubs, pails, and all hooped wooden ware, should be kept out of the sun, and in a cool place, or they will fall to pieces.

Common Mode of Washing.

Assort the clothes, and put them in soak, the night before. Never pour hot water on them, as it sets the dirt. In assorting clothes, put the flannels in one lot,[Pg 286] the colored clothes in another, the coarse white ones in a third, and the fine clothes in a fourth lot. Wash the fine clothes in one tub of suds; and throw them, when wrung, into another. Then wash them, in the second suds, turning them wrong side out. Put them in the boiling-bag, and boil them in strong suds, for half an hour, and not much more. Move them, while boiling, with the clothes-stick. Take them out of the boiling-bag, and put them into a tub of water, and rub the dirtiest places, again, if need be. Throw them into the rinsing-water, and then wring them out, and put them into the blueing-water. Put the articles to be stiffened, into a clothes-basket, by themselves, and, just before hanging out, dip them in starch, clapping it in, so as to have them equally stiff, in all parts. Hang white clothes in the sun, and colored ones, (wrong side out,) in the shade. Fasten them with clothes-pins. Then wash the coarser white articles, in the same manner. Then wash the colored clothes. These must not be soaked, nor have ley or soda put in the water, and they ought not to lie wet long before hanging out, as it injures their colors. Beef's-gall, one spoonful to two pailfuls of suds, improves calicoes. Lastly, wash the flannels, in suds as hot as the hand can bear. Never rub on soap, as this shrinks them in spots. Wring them out of the first suds, and throw them into another tub of hot suds, turning them wrong side out. Then throw them into hot blueing-water. Do not put blueing into suds, as it makes specks in the flannel. Never leave flannels long in water, nor put them in cold or lukewarm water. Before hanging them out, shake and stretch them. Some housekeepers have a close closet, made with slats across the top. On these slats, they put their flannels, when ready to hang out, and then burn brimstone under them, for ten minutes. It is but little trouble, and keeps the flannels as white as new. Wash the colored flannels, and hose, after the white, adding more hot water. Some persons dry woollen hose on stocking-boards, shaped like a foot and leg,[Pg 287] with strings to tie them on the line. This keeps them from shrinking, and makes them look better than if ironed. It is also less work, than to iron them properly.

Bedding should be washed in long days, and in hot weather. Pound blankets in two different tubs or barrels of hot suds, first well mixing the soap and water. Rinse in hot suds; and, after wringing, let two persons shake them thoroughly, and then hang them out. If not dry, at night, fold them, and hang them out the next morning. Bedquilts should be pounded in warm suds; and, after rinsing, be wrung as dry as possible. Bolsters and pillows can be pounded in hot suds, without taking out the feathers, rinsing them in fair water. It is usually best, however, for nice feathers, to take them out, wash them, and dry them on a garret floor. Cotton comforters should have the cases taken off and washed. Wash bedticks, after the feathers are removed, like other things. Empty straw beds once a year.

The following cautions, in regard to calicoes, are useful. Never wash them in very warm water; and change the water, when it appears dingy, or the light parts will look dirty. Never rub on soap; but remove grease with French chalk, starch, magnesia, or Wilmington clay. Make starch for them, with coffee-water, to prevent any whitish appearance. Glue is good for stiffening calicoes. When laid aside, not to be used, all stiffening should be washed out, or they will often be injured. Never let calicoes freeze, in drying. Some persons use bran-water, (four quarts of wheat-bran to two pails of water,) and no soap, for calicoes; washing and rinsing in the bran-water. Potato-water is equally good. Take eight peeled and grated potatoes to one gallon of water.

Soda-Washing.

A very great saving in labor is secured, by *soda-washing*. There have been mistakes made in receipts, and in modes of doing it, which have caused a prejudice against it; but if the soap be rightly made, and rightly[Pg 288] used, *it certainly saves one half the labor and time of ordinary washing.*

Receipt for Soda-Soap.

Take eight pounds of bar-soap, eight pounds of coarse soda, (the sub-carbonate,) ten gallons of soft water, boiled two hours, stirring it often. This is to be cooled, and set away for use. In washing, take a pound of this soap, to the largest pail of water, and heat till it boils. Having previously soaked the white clothes, in *warm*, not *hot*, water, put them in this boiling mixture, and let them boil *one hour and no more*. Take them out, draining them well, and put them in a tub, half full of soft water. Turn them wrong side out; rub the soiled places, till they look clean; then put them into blue rinsing-water, and wring them out. They are then ready to hang out. Some persons use another rinsing-water. The colored clothes and flannels must not be washed in this way. The fine clothes may be first boiled in this water; it may then be used for coarser clothes; and afterward, the brown towels, and other articles of that nature, may be boiled in the same water. After this, the water which remains, is still useful, for washing floors; and then, the suds is a good manure to put around plants.

It is best to prepare, at once, the whole quantity of water to be used. Take out about one third, and set it by; and every time a fresh supply of clothes is put in, use a portion of this, to supply the waste of a former boiling.

Modes of Washing Various Articles.

Brown Linens, or *Muslins*, of tea, drab, or olive, colors, look best, washed in hay-water. Put in hay enough, to color the water like new brown linen. Wash them first in lukewarm, fair water, without soap, (removing grease with French chalk,) then wash and rinse them in the hay-water.

Nankeens look best, washed in suds, with a teacup[Pg 289] of ley added for each pailful. Iron on the wrong side. Soak new nankeens in ley, for one night, and it sets the color perfectly.

Woollen Table-Covers and *Woollen Shawls*, may be washed thus: Remove grease as before directed. If there be stains in the articles, take them out with spirits of hartshorn. Wash the things in two portions of hot suds, made of white soap. Do not wring them, but fold them and press the water out, catching it in a tub, under a table. Shake, stretch, and dry, neither by the sun nor a fire, and do not let them freeze, in drying. Sprinkle them three hours before ironing, and fold and roll them tight. Iron them heavily on the wrong side. *Woollen yarn*, should be washed in very hot water, putting in a tea-cupful of ley, and no soap, to half a pailful of water. Rinse till the water comes off clear.

New Black Worsted and Woollen Hose, should be soaked all night, and washed in hot suds, with beef's-gall, a tablespoonful to half a pail of water. Rinse till no color comes out. Iron on the wrong side.

To Cleanse Gentlemen's Broadcloths. The common mode, is, to shake, and brush the articles, and rip out linings and pockets; then to wash them in strong suds, adding a teacupful of ley, using white soap for light cloth; rolling and then pressing, instead of wringing, them; when dry, sprinkling them, and letting them lie all night; and ironing on the wrong side, or with a thin dark cloth over the article, until *perfectly* dry. But a far better way, which the writer has repeat-edly tried, with unfailing success, is the following: Take one beef's-gall, half a pound of salæratus, and four gallons of warm water. Lay the article on a table, and scour it thoroughly, in every part, with a clothes-brush, dipped in this mixture. The collar of a coat, and the grease-spots, (previously marked by stitches of white thread,) must be repeatedly brushed. Then, take the article, and rinse it up and down in the mixture. Then, rinse it up and down in a tub of soft cold water. Then, without wringing or pressing, hang it to drain and dry.[Pg 290] Fasten a coat up by the collar. When perfectly dry, it is sometimes the case, with coats, that nothing more is needed. In other cases, it is necessary to dampen the parts, which look wrin-kled, with a sponge, and either pull them smooth, with the fingers, or press them with an iron, having a piece of bombazine, or thin woollen cloth, between the iron and the article.

To manufacture Ley, Soap, Starch, and other Articles used in Washing.

To make Ley. Provide a large tub, made of pine or ash, and set it on a form, so high, that a tub can stand under it. Make a hole, an inch in diameter, near the bottom, on one side. Lay bricks, inside, about this hole, and straw over them. To every seven bushels of ashes, add two gallons of unslacked lime, and throw in the ashes and lime in alternate layers. While putting in the ashes and lime, pour on boiling water, using three or four pailfuls. After this, add a pailful of cold soft water, once an hour, till all the ashes appear to be well soaked. Catch the drippings, in a tub, and try its strength with an egg. If the egg rise so as to show a circle as large as a ten cent piece, the strength is right; if it rise higher, the ley must be weakened by water; if not so high, the ashes are not good, and the whole process must be repeated, putting in fresh ashes, and running the weak ley through the new ashes, with some additional water. *Quick-ley* is made by pouring one gallon of boiling soft water on three quarts of ashes, and straining it. Oak ashes are best.

To make Soft-Soap. Save all drippings and fat, melt them, and set them away, in cakes. Some persons keep, for soap-grease, a half barrel, with weak ley in it, and a cover over it. To make soft-soap, take the proportion of one pailful of ley to three pounds of fat. Melt the fat, and pour in the ley, by degrees. Boil it steadily, through the day, till it is ropy. If not boiled enough, on cooling, it will turn to ley and sediment. [Pg 291]While boiling, there should always be a little oil on the surface. If this does not appear, add more grease. If there is too much grease, on cooling, it will rise, and can be skimmed off. Try it, by cooling a small quantity. When it appears like gelly, on becoming cold, it is done. It must then be put in a cool place and often stirred.

To make cold Soft-Soap, melt thirty pounds of grease, put it in a barrel, add four pailfuls of strong ley, and stir it up thoroughly. Then gradually add more ley, till the barrel is nearly full, and the soap looks *about right.*

To make Potash-Soap, melt thirty-nine pounds of grease, and put it in a barrel. Take twenty-nine pounds of light ash-colored potash, (the *reddish*-colored will spoil the soap,) and pour hot water on it; then pour it off into the grease, stirring it well. Continue thus, till all the potash is melted. Add one pailful of cold water, stirring it a great deal, every day, till the barrel be full, and then it is done. This is the cheapest and best kind of soap. It is best to sell ashes and buy potash. The soap is better, if it stand a year before it is used; therefore make two barrels at once.

To make Hard White Soap, take fifteen pounds of lard, or suet; and, when boiling, add, slowly, five gallons of ley, mixed with one gallon of water. Cool a small portion; and, if no grease rise, it is done: if grease do rise, add ley, and boil till no grease rises. Then add three quarts of fine salt, and boil it; if this do not harden well, on cooling, add more salt. Cool it, and if it is to be perfumed, melt it next day, put in the perfume, and then run it in moulds, or cut it in cakes. *Common Hard Soap*, is made in the same way, by using common fat.

To manufacture Starch, cleanse a peck of unground wheat, and soak it, for several days, in soft water. When quite soft, remove the husks, with the hand, and the soft parts will settle. Pour off the water, and replace it, every day, with that which is fresh, stirring it well. When, after stirring and settling, the water is[Pg 292] clear, it is done. Then strain off the water, and dry the starch, for several days, in the sun. If the water be permitted to remain too long, it sours, and the starch is poor. If the starch be not well dried, it grows musty.

CHAPTER XXVII.
ON STARCHING, IRONING, AND CLEANSING.

To prepare Starch. Take four tablespoonfuls of starch; put in as much water; and rub it, till all lumps are removed. Then, add half a cup of cold water. Pour this into a quart of boiling water, and boil it for half an hour, adding a piece of spermaceti, or a lump of salt, or sugar, as large as a hazelnut. Strain it, and put in a very little blueing. Thin it with hot water.

Glue and Gum-Starch. Put a piece of glue, four inches square, into three quarts of water, boil it, and keep it in a bottle, corked up. Dissolve four ounces of gum Arabic, in a quart of hot water, and set it away, in a bottle, corked. Use the glue for calicoes, and the gum for silks and muslins, both to be mixed with water, at discretion.

Beef's-Gall. Send a junk-bottle to the butcher, and have several gall-bladders emptied into it. Keep it salted, and in a cool place. Some persons perfume it; but fresh air removes the unpleasant smell which it gives, when used for clothes.

Directions for Starching Muslins and Laces.

Many ladies clap muslins, then dry them, and afterwards sprinkle them. This saves time. Others clap them, till nearly dry, then fold and cover, and then iron them. Iron wrought muslins on soft flannel, and on the wrong side.

To do up Laces, nicely, sew a clean piece of muslin around a long bottle, and roll the lace on it; pulling[Pg 293] out the edge, and rolling it so that the edge will turn in, and be covered, as you roll. Fill the bottle with water, and then boil it, for an hour, in a suds made with white soap. Rinse it in fair water, a little blued; dry it in the sun; and, if any stiffening is wished, use thin starch, or gum Arabic. When dry, fold and press it, between white papers, in a large book. It improves the lace, to wet it with sweet-oil, after it is rolled on the bottle, and before boiling in the suds. *Blond laces* can be whitened, by rolling them on a bottle, in this way, and then setting the bottle in the sun, in a dish of cold suds made with white soap, wetting it thoroughly, and changing the suds, every day. Do this, for a week or more; then rinse, in fair water; dry it on the bottle, in the sun; and stiffen it with white gum Arabic. Lay it away in loose folds. *Lace veils* can be whitened, by laying them in flat dishes, in suds made with white soap; then rinsing, and stiffening them with gum Arabic, stretching them, and pinning them on a sheet, to dry.

ON IRONING.

Articles to be provided for Ironing.

A settee, or settle, made so that it can be used for an ironing-table, is a great convenience. It may be made of pine, and of the following dimensions: length, five feet and six inches; width of the seat, one foot and nine inches; height of the seat, one foot and three inches; height of the sides, (or arms of the seat,) two feet and four inches; height of the back, five feet and three inches. The back should be made with hinges, of the height of the sides or arms, so that it can be turned down, and rest on them, and thus become an ironing-table. The back is to be fastened up, behind, with long iron hooks and staples. The seat should be made with two lids, opening into two boxes, or partitions, in one of which, can be kept the ironing-sheets and holders, and in the other, the other articles used in ironing. It can be stained of a cherry-color; put on casters, so as to move easily;[Pg 294] and be provided with two cushions, stuffed with hay and covered with dark woollen. It thus serves as a comfortable seat, for Winter, protecting the back from cold.

Where a settee, of this description, is not provided, a large ironing-board, made so as not to warp, should be kept, and used only for this purpose, to be laid, when used, on a table. Provide, also, the following articles: A woollen ironing-blanket, and a linen or cotton sheet, to spread over it; a large fire, of charcoal and hard wood, (unless furnaces or stoves are used;) a hearth, free from cinders and ashes, a piece of sheet-iron, in front of the fire, on which to set the irons, while heating; (this last saves many black spots from careless ironers;) three or four holders, made of woollen, and covered with old silk, as these do not easily take fire; two iron rings, or iron-stands, on which to set the irons, and small pieces of board to put under them, to prevent scorching the sheet; linen or cotton wipers; and a piece of beeswax, to rub on the irons when they are smoked. There should be, at least, three irons for each person ironing, and a small and large clothes-frame, on which to air the fine and coarse clothes.

A bosom-board, on which to iron shirt-bosoms, should be made, one foot and a half long, and nine inches wide, and covered with white flannel. A skirt-board on which to iron frock-skirts, should be made, five feet long, and two feet wide at one end, tapering to one foot and three inches wide, at the other end. This should be covered with flannel; and will save much trouble, in ironing nice dresses. The large end may be put on the table, and the other, on the back of a chair. Both these boards should have cotton covers, made to fit them; and these should be changed and washed, when dirty. These boards are often useful, when articles are to be ironed or pressed, in a chamber or parlor. Provide, also, a press-board, for broadcloth, two feet long, and four inches wide at one end, tapering to three inches wide, at the other.

[Pg 295]

A fluting-iron, called, also, a patent Italian iron, saves much labor, in ironing ruffles neatly. A crimping-iron, will crimp ruffles beautifully, with very little time or trouble. Care must be used, with the latter, or it will cut the ruffles. A trial should be made, with old muslins; and, when the iron is screwed in the right place, it must be so kept, and not altered without leave from the housekeeper. If the lady of the house will provide all these articles, see that the fires are properly made, the ironing-sheets evenly put on and properly pinned, the clothes-frames dusted, and all articles kept in their places, she will do much towards securing good ironing.

On Sprinkling, Folding, and Ironing.

Wipe the dust from the ironing-board, and lay it down, to receive the clothes, which should be sprinkled with clear water, and laid in separate piles, one of colored, one of common, and one of fine articles, and one of flannels. Fold the fine things, and roll them in a towel, and then fold the rest, turning them all right side outward.

The colored clothes should be laid separate from the rest, and ought not to lie long damp, as it injures the colors. The sheets and table linen should be shaken, stretched, and folded, by two persons. Iron lace and needlework on the wrong side, and carry them away, as soon as dry. Iron calicoes with irons which are not very hot, and generally on the right side, as they thus keep clean for a longer time. In ironing a frock, first do the waist, then the sleeves, then the skirt. Keep the skirt rolled, while ironing the other parts, and set a chair, to hold the sleeves, while ironing the skirt, unless a skirt-board be used. In ironing a shirt, first do the back, then the sleeves, then the collar and bosom, and then the front. Iron silk on the wrong side, when quite damp, with an iron which is not very hot. Light colors are apt to change and fade. Iron velvet, by turning up the face of the iron, and after dampening the wrong side of the velvet, draw it over the face of the iron, holding it straight, and not biased.

[Pg 296]

CHAPTER XXVIII.
ON WHITENING, CLEANSING, AND DYEING.

To Whiten Articles, and Remove Stains from them.

Wet white clothes in suds, and lay them on the grass, in the sun. Lay muslins in suds made with white soap, in a flat dish; set this in the sun, changing the suds, every day. Whiten tow-cloth, or brown linen, by keeping it in ley, through the night, laying it out in the sun, and wetting it with fair water, as fast as it dries.

Scorched articles can often be whitened again, by laying them in the sun, wet with suds. Where this does not answer, put a pound of white soap in a gallon of milk, and boil the article in it. Another method, is, to chop and extract the juice from two onions, and boil this with half a pint of vinegar, an ounce of white soap, and two ounces of fuller's earth. Spread this, when cool, on the scorched part, and, when dry, wash it off, in fair water. *Mildew* may be removed, by dipping the article in sour buttermilk, laying it in the sun, and, after it is white, rinsing it in fair water. Soap and chalk are also good; also, soap and starch, adding half as much salt as there is starch, together with the juice of a lemon. Stains in linen can often be removed, by rubbing on soft soap, then putting on a starch paste, and drying in the sun, renewing it several times. Wash off all the soap and starch, in cold, fair water.

Mixtures for Removing Stains and Grease.

Stain-Mixture. Half an ounce of oxalic acid, in a pint of soft water. This can be kept in a corked bottle, and is infallible in removing iron-rust, and ink-stains. It is very poisonous. The article must be spread with this mixture over the steam of hot water, and wet several times. This will also remove indelible ink.[Pg 297] The article must be washed, or the mixture will injure it.

Another Stain-Mixture is made, by mixing one ounce of sal ammoniac, one ounce of salt of tartar, and one pint of soft water.

To remove Grease. Mix four ounces of fuller's earth, half an ounce of pearlash, and lemon-juice enough to make a stiff paste, which can be dried in balls, and kept for use. Wet the greased spot with cold water, rub it with the ball, dry it, and then rinse it with fair cold water. This is for *white* articles. For silks, and worsteds, use French chalk, which can be procured of the apothecaries. That which is soft and white, is best. Scrape it on the greased spot, and let it lie for a day and night. Then renew it, till the spot disappears. Wilmington clay-balls, are equally good. Ink-spots can often be removed from white clothes, by rubbing on common tallow, leaving it for a day or two, and then washing, as usual. Grease can be taken out of wall-paper, by making a paste of potter's clay, water and ox-gall, and spreading it on the paper. When dry, renew it, till the spot disappears.

Stains on floors, from *soot,* or *stove-pipes,* can be removed, by washing the spot in sulphuric acid and water. Stains, in colored silk dresses, can often be removed, by pure water. Those made by acids, tea, wine, and fruits, can often be removed, by spirits of hartshorn, diluted with an equal quantity of water. Sometimes, it must be repeated, several times.

Tar, Pitch, and *Turpentine,* can be removed, by putting the spot in sweet-oil, or by spreading tallow on it, and letting it remain for twenty-four hours. Then, if the article be linen or cotton, wash it, as usual; if it be silk or worsted, rub it with ether, or spirits of wine.

Lamp-Oil can be removed, from floors, carpets, and other articles, by spreading upon the stain a paste, made of fuller's earth or potter's clay, and renewing it, when dry, till the stain is removed. If gall be put into the paste, it will preserve the colors from injury. When[Pg 298] the stain has been removed, carefully brush off the paste, with a soft brush.

Oil-Paint can be removed, by rubbing it with *very pure* spirits of turpentine. The impure spirit leaves a grease-spot. *Wax* can be removed, by scraping it off, and then holding a red-hot poker near the spot. *Spermaceti* may be removed by scraping it off, then putting a paper over the spot, and applying a warm iron. If this does not answer, rub on spirits of wine.

Ink-Stains, in carpets and woollen table-covers, can be removed, by washing the spot in a liquid, composed of one teaspoonful of oxalic acid dissolved in a teacupful of warm (not hot) water, and then rinsing in cold water.

Stains on Varnished Articles, which are caused by cups of hot water, can be removed, by rubbing them with lamp-oil, and then with alcohol. Ink-stains can be taken out of mahogany, by one teaspoonful of oil of vitriol mixed with one tablespoonful of water, or by oxalic acid and water. These must be brushed over quickly, and then washed off with milk.

Modes of Cleansing Various Articles.

Silk Handkerchiefs and *Ribands* can be cleansed, by using French chalk to take out the grease, and then sponging them, on both sides, with lukewarm fair water. Stiffen them with gum Arabic, and press them between white paper, with an iron not very hot. A tablespoonful of spirits of wine to three quarts of water, improves it.

Silk Hose, or *Silk Gloves*, should be washed in warm suds made with white soap, and rinsed in cold water; they should then be stretched and rubbed, with a hard-rolled flannel, till they are quite dry. Ironing them, very much injures their looks. *Washleather* articles should have the grease removed from them, by French chalk, or magnesia; they should then be washed in warm suds, and rinsed in cold water. *White Kid Gloves* should have the grease removed from them, as[Pg 299] above directed. They should then be brushed, with a soft brush, and a mixture of fuller's earth and magnesia. In an hour after, rub them with flannel, dipped in bran and powdered whiting. *Colored or Hoskin's gloves* can be cleansed, very nicely, by *pure* spirits of turpentine, put on with a woollen cloth, and rubbed from wrist to fingers. Hang them for several days in the air, and all the unpleasant smell will be removed. *Gentlemen's white gloves* should be washed with a sponge, in white-soapsuds; then wiped, and dried on the hands. *Swan's-down tippets, and capes*, should be washed in white-soapsuds, squeezing, and not rubbing them; then rinse them in two waters, and shake and stretch them while drying. *Ostrich feathers* can also be thus washed. Stiffen them, with starch, wet in cold water and not boiled. Shake them in the air, till nearly dry, then hold them before the fire, and curl them with dull scissors, giving each fibre a twitch, turning it inward, and holding it so for a moment.

Straw and Leghorn Hats, can be cleansed, by simply washing them in white-soapsuds. Remove grease, by French chalk, and stains, by diluted oxalic acid, or cream of tartar. The oxalic acid is best, but must be instantly washed off. *To whiten them*, drive nails in a barrel, near its bottom, so that cords can be stretched across. On these cords, tie the bonnet, wet with suds, (having first removed the grease, stains, and dirt.) Then invert the barrel, over a dish of coals, on which roll brimstone is slowly burning. Put a chip under one side of the barrel, to admit the air. Continue this, till the bonnet is white; then hang it in the air, (when the weather is not damp,) till the smell is removed. Then stiffen it with a solution of isinglass or gum Arabic, put on the inside, with a sponge. Press the crown, on a block, and the rest on a board, on the right side, putting muslin between the iron and straw, and pressing hard. Be careful not to make it too stiff. First, stiffen a small piece, for trial.

[Pg 300]

ON COLORING.

Precautions and Preparations.

All the articles must be entirely free from grease or oil, and also, in most cases, from soapsuds. Make light dyes in brass, and dark ones in iron, vessels. Always wet the articles, in fair water, before dyeing. Always carefully strain the dye. If the color be too light, dry and then dip the article again. Stir the article well in the dye, lifting it up often. Remove any previous color, by boiling in suds, or, what is better, in the soda mixture used for washing.

Pink Dye. Buy a saucer of carmine, at an apothecary's. With it, you will find directions for its use. This is cheap, easy to use, and beautiful. *Balm blossoms* and *Bergamot blossoms*, with a little cream of tartar in the water, make a pretty pink.

316

Red Dye. Take half a pound of wheat bran, three ounces of powdered alum, and two gallons of soft water. Boil these in a brass vessel, and add an ounce of cream of tartar, and an ounce of cochineal, tied up together in a bag. Boil the mixture for fifteen minutes, then strain it, and dip the articles. Brazil wood, set with alum, makes another red dye.

Yellow Dye. Fustic, turmeric powder, saffron, barberry-bush, peach-leaves, or marigold flowers, make a yellow dye. Set the dye with alum, putting a piece the size of a large hazelnut to each quart of water.

Light Blue Dye, for silks and woollens, is made with the 'blue composition,' to be procured of the hat-makers; fifteen drops to a quart of water. Articles dipped in this, must be thoroughly rinsed. For a *dark blue*, boil four ounces of copperas in two gallons of water. Dip the articles in this, and then in a strong decoction of logwood, boiled and strained. Then wash them thoroughly in soapsuds.

Green Dye. First color the article yellow; and[Pg 301] then, if it be silk or woollen, dip it in 'blue composition.' Instead of ironing, rub it with flannel, while drying.

Salmon Color is made by boiling arnotto or anotta in soapsuds.

Buff Color is made by putting one teacupful of potash, tied in a bag, in two gallons of hot (not boiling) water, and adding an ounce of arnotto, also in a bag, keeping it in for half an hour. First, wet the article in strong potash-water. Dry and then rinse in soapsuds. Birch bark and alum also make a buff. Black alder, set with ley, makes an orange color.

Dove and Slate Colors, of all shades, are made by boiling, in an iron vessel, a teacupful of black tea, with a teaspoonful of copperas. Dilute this, till you get the shade wanted. Purple sugar-paper, boiled, and set with alum, makes a similar color.

Brown Dye. Boil half a pound of camwood (in a bag) in two gallons of water, for fifteen minutes. Wet the articles, and boil them for a few minutes in the dye. White-walnut bark, the bark of sour sumach, or of white maple, set with alum, make a brown color.

Black Dye. Let one pound of chopped logwood remain all night in one gallon of vinegar. Then boil them, and put in a piece of copperas, as large as a hen's egg. Wet the articles in warm water, and put them in the dye, boiling and stirring them for fifteen minutes. Dry them, then wet them in warm water, and dip them again. Repeat the process, till the articles are black enough. Wash them in suds, and rinse them till the water comes off clear. Iron nails, boiled in vinegar, make a black dye, which is good for restoring rusty black silks.

Olive Color. Boil fustic and yellow-oak bark together. The more fustic, the brighter the olive; the more oak bark, the darker the shade. Set the light shade with a few drops of oil of vitriol, and the dark shade with copperas.

[Pg 302]

CHAPTER XXIX.
ON THE CARE OF PARLORS.

In selecting the furniture of parlors, some reference should be had to correspondence of shades and colors. Curtains should be darker than the walls; and, if the walls and carpets be light, the chairs should be dark, and *vice versa*. Pictures always look best on light walls.

In selecting carpets, for rooms much used, it is poor economy to buy cheap ones. *Ingrain* carpets, of close texture, and the *three-ply* carpets, are best for common use. *Brussels* carpets do not wear so long as the three-ply ones, because they cannot be turned. *Wilton* carpets wear badly, and *Venetians* are good only for halls and stairs.

In selecting colors, avoid those in which there are any black threads; as they are always rotten. The most tasteful carpets, are those, which are made of various shades of the same color, or of all shades of only two colors; such as brown and yellow, or blue and buff, or salmon and green, or all shades of green, or of brown. All very dark shades should be brown or green, but not black.

In laying down carpets, it is a bad practice to put straw under them, as this makes them wear out in spots. Straw matting, laid under carpets, makes them last much longer, as it is smooth and even, and the dust sifts through it. In buying carpets, always get a few yards over, to allow for waste in matching figures.

In cutting carpets, make them three or four inches shorter than the room, to allow for stretching. Begin to cut *in the middle* of a figure, and it will usually match better. Many carpets match in two different ways, and care must be taken to get the right one. Sew a carpet on the wrong side, with double waxed[Pg 303] thread, and with the *ball-stitch*. This is done by taking a stitch on the breadth next you, pointing the needle towards you; and then taking a stitch on the other breadth, pointing the needle from you. Draw the thread tightly, but not so as to pucker. In fitting a breadth to the hearth, cut slits in the right place, and turn the piece under. Bind *the whole* of the carpet, with carpet-binding, nail it with tacks, having bits of leather under the heads. To stretch the carpet, use a carpet-fork, which is a long stick, ending with notched tin, like saw-teeth. This is put in the edge of the carpet, and pushed by one person, while the nail is driven by another. Cover blocks, or bricks, with carpeting, like that of the room, and put them behind tables, doors, sofas, &c., to preserve the walls from injury, by knocking, or by the dusting-cloth.

Cheap footstools, made of a square plank, covered with tow-cloth, stuffed, and then covered with carpeting, with worsted handles, look very well. Sweep carpets as seldom as possible, as it wears them out. To shake them often, is good economy. In cleaning carpets, use damp tea leaves, or wet Indian meal, throwing it about, and rubbing it over with the broom. The latter, is very good for cleansing carpets made dingy by coal-dust. In brushing carpets in ordinary use, it will be found very convenient to use a large flat dust-pan, with a perpendicular handle a yard high, put on so that the pan will stand alone. This can be carried about, and used without stooping, brushing dust into it with a common broom. The pan must be very large, or it will be upset.

When carpets are taken up, they should be hung on a line, or laid on long grass, and whipped, first on one side, and then on the other, with pliant whips. If laid aside, they should be sewed up tight, in linen, having snuff or tobacco put along all the crevices where moths could enter. Shaking pepper, from a pepper-box, round the edge of the floor, under a carpet, prevents the access of moths.

[Pg 304]

Carpets can be best washed on the floor, thus: First shake them; and then, after cleaning the floor, stretch and nail them upon it. Then scrub them in cold soapsuds, having half a teacupful of ox-gall to a bucket of water. Then wash off the suds, with a cloth, in fair water. Set open the doors and windows, for two days or more. Imperial Brussels, Venetian, ingrain, and three-ply, carpets, can be washed thus; but Wilton, and other plush-carpets, cannot. Before washing them, take out grease, with a paste, made of potter's clay, ox-gall, and water.

Straw matting is best for chambers and Summer parlors. The checked, of two colors, is not so good to wear. The best, is the cheapest in the end. When washed, it should be done with salt water, wiping it dry; but frequent washing injures it. Bind matting with cotton binding. Sew breadths together like carpeting. In joining the ends of pieces, ravel out a part, and tie the threads together, turning under a little of each piece, and then, laying the ends close, nail them down, with nails having kid under their heads.

In hanging pictures, put them so that the lower part shall be opposite the eye. Cleanse the glass of pictures with whiting, as water endangers the pictures. Gilt frames can be much better preserved by putting on a coat of copal varnish, which, with proper brushes, can be bought of carriage or cabinet-makers. When dry, it can be washed with fair water. Wash the brush in spirits of turpentine.

Curtains, ottomans, and sofas covered with worsted, can be cleansed, by wheat-bran, rubbed on with flannel. Dust Venetian blinds with feather brushes. Buy light-colored ones, as the green are going out of fashion. Strips of linen or cotton, on rollers and pulleys, are much in use, to shut out the sun from curtains and carpets. Paper curtains, pasted on old cotton, are good for chambers. Put them on rollers, having cords nailed to them, so that when the curtain falls, the cord will be wound up. Then, by pulling the cord, the curtain will be rolled up.

[Pg 305]

Mahogany furniture should be made in the Spring, and stand some months before it is used, or it will shrink and warp. Varnished furniture should be rubbed only with silk, except occasionally, when a little sweet-oil should be rubbed over, and wiped off carefully. For unvarnished furniture, use beeswax, a little softened with sweet-oil; rub it in with a hard brush, and polish with woollen and silk rags. Some persons rub in linseed-oil; others mix beeswax with a little spirits of turpentine and rosin, making it so that it can be put on with a sponge, and wiped off with a soft rag. Others, keep in a bottle the following mixture; two ounces of spirits of turpentine, four tablespoonfuls of sweet-oil, and one quart of milk. This is applied with a sponge, and wiped off with a linen rag.

Hearths and jambs, of brick, look best painted over with black-lead, mixed with soft-soap. Wash the bricks which are nearest the fire with redding and milk, using a painter's brush. A sheet of zinc, covering the whole hearth, is cheap, saves work, and looks very well. A tinman can fit it properly.

Stone hearths should be rubbed with a paste of powdered stone, (to be procured of the stonecutters,) and then brushed with a stiff brush. Kitchen-hearths, of stone, are improved by rubbing in lamp-oil.

Stains can be removed from marble, by oxalic acid and water, or oil of vitriol and water, left on fifteen minutes, and then rubbed dry. Gray marble is improved by linseed-oil. Grease can be taken from marble, by ox-gall and potter's clay wet with soapsuds, (a gill of each.) It is better to add, also, a gill of spirits of turpentine. It improves the looks of marble, to cover it with this mixture, leaving it two days, and then rubbing it off.

Unless a parlor is in constant use, it is best to sweep it only once a week, and at other times use a whisk-broom and dust-pan. When a parlor with handsome furniture is to be swept, cover the sofas, centre table,[Pg 306] piano, books, and mantelpiece, with old cottons, kept for the purpose. Remove the rugs, and shake them, and clean the jambs, hearth, and fire-furniture. Then sweep the room, moving every article. Dust the furniture, with a dust-brush and a piece of old silk. A painter's brush should be kept, to remove dust from ledges and crevices. The dust-cloths should be often shaken and washed, or else they will soil the walls and furniture when they are used. Dust ornaments, and fine books, with feather brushes, kept for the purpose.

CHAPTER XXX.
ON THE CARE OF BREAKFAST AND DINING-ROOMS.

An eating-room should have in it a large closet, with drawers and shelves, in which should be kept all the articles used at meals. This, if possible, should communicate with the kitchen, by a sliding window, or by a door, and have in it a window, and also a small sink, made of marble or lined with zinc, which will be a great convenience for washing nice articles. If there be a dumb-waiter, it is best to have it connected with such a closet. It may be so contrived, that, when it is down, it shall form part of the closet floor.

A table-rug, or crumb-cloth, is useful to save carpets from injury. Bocking, or baize, is best. Always spread the same side up, or the carpet will be soiled by the rug. Table-mats are needful, to prevent injury to the table from the warm dishes. Teacup-mats, or small plates, are useful to save the table-cloths from dripping tea or coffee. Butter-knives, for the butter-plate, and salt-spoons, for salt-dishes, are designed to prevent those disgusting marks which are made, when persons use their own knives, to take salt or butter. A sugar-spoon should be kept in or by the sugar-dish, for the same purpose. Table-napkins, of diaper, are often laid[Pg 307] by each person's plate, for use during the meal, to save the tablecloth and pocket-handkerchief. To preserve the same napkin for the same person, each member of the family has a given number, and the napkins are numbered to correspond, or else are slipped into ivory rings, which are numbered. A stranger has a clean one, at each meal. Tablecloths should be well starched, and ironed on the right side, and always, when taken off, folded in the ironed creases. *Doilies* are colored napkins, which, when fruit is offered, should always be furnished, to prevent a person from staining a nice handkerchief, or permitting the fruit-juice to dry on the fingers.

Casters and salt-stands should be put in order, every morning, when washing the breakfast things. Always, if possible, provide *fine* and *dry* table-salt, as many persons are much disgusted with that which is dark, damp, and coarse. Be careful to keep salad-oil closely corked, or it will grow rancid. Never leave the salt-spoons in the salt, nor the mustard-spoon in the mustard, as they are thereby injured. Wipe them, immediately after the meal.

For table-furniture, French china is deemed the nicest, but it is liable to the objection of having plates, so made, that salt, butter, and similar articles, will not lodge on the edge, but slip into the centre. Select knives and forks, which have weights in the handles, so that, when laid down, they will not touch the table. Those with rivetted handles last longer than any others. Horn handles (except buckhorn) are very poor. The best are cheapest in the end. Knives should be sharpened once a month, unless they are kept sharp by the mode of scouring.

On Setting Tables.

Neat housekeepers observe the manner in which a table is set more than any thing else; and to a person of good taste, few things are more annoying, than to see the table placed askew; the tablecloth soiled, rumpled, and put on awry; the plates, knives, and dishes thrown[Pg 308] about, without any order; the pitchers soiled on the outside, and sometimes within; the tumblers dim; the caster out of order; the butter pitched on the plate, without any symmetry; the salt coarse, damp, and dark; the bread cut in a mixture of junks and slices; the dishes of food set on at random, and without mats; the knives dark or rusty, and their handles greasy; the tea-furniture all out of order, and every thing in similar style. And yet, many of these negligences will be met with, at the tables of persons who call themselves well bred, and who have wealth enough to make much outside show. One reason for this, is, the great difficulty of finding domestics, who will attend to these things in a proper manner, and who, after they have been repeatedly instructed, will not neglect nor forget what has been said to them. The writer has known cases, where much has been gained by placing the following rules in plain sight, in the place where the articles for setting tables are kept.

Rules for setting a Table.

1. Lay the rug square with the room, and also smooth and even; then set the table also square with the room, and see that the *legs* are in the right position to support the leaves.

2. Lay the tablecloth square with the table, *right side up*, smooth, and even.

3. Put on the teatray (for breakfast or tea) square with the table; set the cups and saucers at the front side of the teatray, and the sugar, slop-bowls, and cream-cup, at the back side. Lay the sugar-spoon or tongs on the sugar-bowl.

4. Lay the plates around the table, at equal intervals, and the knives and forks at regular distances, each in the same particular manner, with a cup-mat, or cup-plate, to each, and a napkin at the right side of each person.

5. If meat be used, set the caster and salt-cellars in the centre of the table; then lay mats for the dishes,[Pg 309] and place the carving-knife and fork and steel by the master of the house. Set the butter on two plates, one on either side, with a butter-knife by each.

6. Set the tea or coffee-pot on a mat, at the right hand of the teatray, (if there be not room upon it.) Then place the chairs around the table, and call the family.

For Dinner.

1. Place the rug, table, tablecloth, plates, knives and forks, and napkins, as before directed, with a tumbler by each plate. In cold weather, set the plates where they will be warmed.

2. Put the caster in the centre, and the salt-stands at two oblique corners, of the table, the latter between two large spoons crossed. If more spoons be needed, lay them on each side of the caster, crossed. Set the pitcher on a mat, either at a side-table, or, when there is no waiter, on the dining-table. Water looks best in glass decanters.

3. Set the bread on the table, when there is no waiter. Some take a fork, and lay a piece on the napkin or tumbler by each plate. Others keep it in a tray, covered with a white napkin to keep off flies. Bread for dinner is often cut in small junks, and not in slices.

4. Set the principal dish before the master of the house, and the other dishes in a regular manner. Put the carving-knife, fork, and steel, by the principal dish, and also a knife-rest, if one be used.

5. Put a small knife and fork by the pickles, and also by any other dishes which need them. Then place the chairs.

On Waiting at Table.

A domestic, who waits on the table, should be required to keep the hair and hands in neat order, and have on a clean apron. A small teatray should be used to carry cups and plates. The waiter should announce the meal (when ready) to the mistress of the family, then stand[Pg 310] by the eating-room door, till all are in, then close the door, and step to the left side of the lady of the house. When all are seated, the waiter should remove the covers, taking care first to invert them, so as not to drop the steam on the table-cloth or guests. In presenting articles, go to the left side of the person. In pouring water never entirely fill the tumbler. The waiter should notice when bread or water is wanting, and hand it without being called. When plates are changed, be careful not to drop knives or forks. Brush off crumbs, with a crumb-brush, into a small waiter.

When there is no domestic waiter, a light table should be set at the left side of the mistress of the house, on which the bread, water, and other articles not in immediate use, can be placed.

On Carving and Helping at Table.

It is considered an accomplishment for a lady to know how to carve well, at her own table. It is not proper to stand in carving. The carving-knife should be sharp and thin. To carve fowls, (which should always be laid with the breast uppermost,) place the fork in the breast, and take off the wings and legs without turning the fowl; then cut out the merry thought, cut slices from the breast, take out the collar bone, cut off the side pieces, and then cut the carcass in two. Divide the joints in the leg of a turkey.

In helping the guests, when no choice is expressed, give a piece of both the white and dark meat, with some of the stuffing. Inquire whether the guest will be helped to each kind of vegetable, and put the gravy on the plate, and not on any article of food.

In carving a sirloin, cut thin slices from the side next to you, (it must be put on the dish with the tenderloin underneath;) then turn it, and cut from the tenderloin Help the guest to both kinds.

In carving a leg of mutton, or a ham, begin by cutting across the middle, to the bone. Cut a tongue across, and not lengthwise, and help from the middle part.

[Pg 311]

Carve a forequarter of lamb, by separating the shoulder from the ribs, and then dividing the ribs. To carve a loin of veal, begin at the smaller end and separate the ribs. Help each one to a piece of the kidney and its fat. Carve pork and mutton in the same way.

To carve a fillet of veal, begin at the top, and help to the stuffing with each slice. In a breast of veal, separate the breast and brisket, and then cut them up, asking which part is preferred. In carving a pig, it is customary to divide it, and take off the head, before it comes to the table; as, to many persons, the head is very revolting. Cut off the limbs, and divide the ribs. In carving venison, make a deep incision down to the bone, to let out the juices; then turn the broad end of the haunch towards you, cutting deep, in thin slices. For a saddle of venison, cut from the tail towards the other end, on each side, in thin slices. Warm plates are very necessary, with venison and mutton, and in Winter, are desirable for all meats.

CHAPTER XXXI.
ON THE CARE OF CHAMBERS AND BEDROOMS.

Every mistress of a family should see, not only that all sleeping-rooms in her house *can be* well ventilated at night, but that they actually are so. Where there is no open fireplace to admit the pure air from the exterior, a door should be left open into an entry, or room where fresh air is admitted; or else a small opening should be made in a window, taking care not to allow a draught of air to cross the bed. The debility of childhood, the lassitude of domestics, and the ill-health of families, are often caused by neglecting to provide a supply of pure air. Straw matting is best for a chamber carpet, and strips of woollen carpeting may be laid[Pg 312] by the side of the bed. Where chambers have no closets, a *wardrobe* is indispensable. This is a moveable closet, with doors, divided, by a perpendicular partition, into two apartments. In one division, rows of hooks are placed, on which to hang dresses. The other division is fitted up with shelves, for other uses. Some are made with drawers at the bottom for shoes, and such like articles. A low square box, set on casters, with a cushion on the top, and a drawer on one side to put shoes in, is a great convenience in dressing the feet. An old champaigne basket, fitted up with a cushion on the lid, and a valance fastened to it to cover the sides, can be used for the same purpose.

A comfortable couch, for chambers and sitting-rooms, can be made by a common carpenter, at a small expense. Have a frame made (like the annexed engraving, Fig. 38,) of common stuff, six feet long, twenty-eight inches wide, and twelve inches high. It must be made thus low, because the casters and cushions will raise it several inches. Have the sloping side-piece, *a*, and head-piece, *b*, sawed out of a board; nail brown linen on them, and stuff them with soft hay or hair. Let these be screwed to the frame, and covered with furniture patch. Then let slats be nailed across the bottom, as at *c, c*, four inches apart. This will cost two or three dollars. Then make a thick cushion, of hay or straw, with side strips, like a mattress, and lay this for the under-cushion. To put over this, make a thinner cushion, of hair, cover it with furniture-calico, and fasten to it a valance reaching to the floor. Then make two square pillows, and cover them with calico, like the rest. Both the cushions should be stitched through like mattresses.

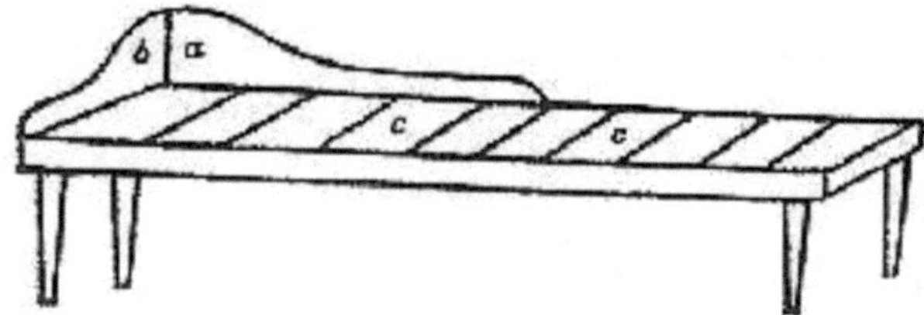

Fig. 38.

[Pg 313]

The writer has seen a couch of this kind, in a common parlor, which cost less than eight dollars, was much admired, and was a constant comfort to the feeble mother, as well as many other members of the family.

Another convenience, for a room where sewing is done in Summer, is a fancy-jar, set in one corner, to receive clippings, and any other rubbish. It can be covered with prints, or paintings, and varnished; and then looks very prettily.

The trunks in a chamber can be improved in looks and comfort, by making cushions of the same size and shape, stuffed with hay and covered with chintz, with a frill reaching nearly to the floor.

Every bedchamber should have a washstand, bowl, pitcher, and tumbler, with a washbucket under the stand, to receive slops. A light screen, made like a clothes-frame, and covered with paper or chintz, should be furnished for bedrooms occupied by two persons, so that ablutions can be performed in privacy. It can be ornamented, so as to look well anywhere. A little frame, or towel-horse, by the washstand, on which to dry towels, is a convenience. A washstand should be furnished with a sponge or washcloth, and a small towel, for wiping the basin after using it. This should be hung on the washstand or towel-horse, for constant use. A soap-dish, and a dish for toothbrushes, are neat and convenient, and each person should be furnished with two towels; one for the feet, and one for other purposes.

It is in good taste to have the curtains, bedquilt, valance, and window-curtains, of similar materials. In making featherbeds, side-pieces should be put in, like those of mattresses, and the bed should be well filled, so that a person will not be buried in a hollow, which is not healthful, save in extremely cold weather. Featherbeds should never be used, except in cold weather. At other times, a thin mattress of hair, cotton and moss, or straw, should be put over them. A simple strip of[Pg 314] broad straw matting, spread over a feather-bed, answers the same purpose. Nothing is more debilitating, than, in warm weather, to sleep with a featherbed pressing round the greater part of the body. Pillows stuffed with papers an inch square, are good for Summer, especially for young children, whose heads should be kept cool. The cheapest and best covering of a bed, for Winter, is a *cotton comforter*, made to contain three or four pounds of cotton, laid in batts or sheets, between covers tacked together at regular intervals. They should be three yards square, and less cotton should be put at the sides that are tucked in. It is better to have two thin comforters, to each bed, than one thick one; as then the covering can be regulated according to the weather.

Few domestics will make a bed properly, without much attention from the mistress of the family. The following directions should be given to those who do this work.

Open the windows, and lay off the bed-covering, on two chairs, at the foot of the bed. After the bed is well aired, shake the feathers, from each corner to the middle; then take up the middle, and shake it well, and turn the bed over. Then push the feathers in place, making the head higher than the foot, and the sides even, and as high as the middle part. Then put on the bolster and the under sheet, so that the wrong side of the sheet shall go next the bed, and the *marking* come at the head, tucking in all around. Then put on the pillows, even, so that the open ends shall come to the sides of the bed, and then spread on the upper sheet, so that the wrong side shall be next the blankets, and the marked end at the head. This arrangement of sheets is to prevent the part where the feet lie from being reversed, so as to come to the face, and also to prevent the parts soiled by the body from coming to the bedtick and blankets. Then put on the other covering, except the outer one, tucking in all around, and then turn over the upper sheet, at the head, so as[Pg 315] to show a part of the pillows. When the pillow-cases are clean and smooth, they look best outside of the cover, but not otherwise. Then draw the hand along the side of the pillows, to make an even indentation, and then smooth and shape the whole outside. A nice housekeeper always notices the manner in which a bed is made; and in some parts of the Country, it is rare to see this work properly performed.

The writer would here urge every mistress of a family, who keeps more than one domestic, to provide them with single beds, that they may not be obliged to sleep with all the changing domestics, who come and go so often. Where the room is too small for two beds, a narrow trucklebed under another, will answer. Domestics should be furnished with washing conveniences in their chambers, and be encouraged to keep their persons and rooms neat and in order.

On Packing and Storing Articles.

Fold a gentleman's coat, thus:—Lay it on a table or bed, the inside downward, and unroll the collar. Double each sleeve once, making the crease at the elbow, and laying them so as to make the fewest wrinkles, and parallel with the skirts. Turn the fronts over the back and sleeves, and then turn up the skirts, making all as smooth as possible.

Fold a shirt, thus:—One that has a bosom-piece inserted, lay on a bed, bosom downward. Fold each sleeve twice, and lay it parallel with the sides of the shirt. Turn the two sides, with the sleeves, over the middle part, and then turn up the bottom, with two folds. This makes the collar and bosom lie, unpressed, on the outside.

Fold a frock thus:—Lay its front downward, so as to make the first creases in folding come in the side breadths. To do this, find the middle of the side breadths by first putting the middle of the front and back breadths together. Next, fold over the side creases so as just to meet the slit behind. Then fold the skirt again, so as to make the backs[Pg 316] lie together within and the fronts without. Then arrange the waist and sleeves, and fold the skirt around them.

In packing trunks, for travelling, put all heavy articles at the bottom, covered with paper, which should not be printed, as the ink rubs off. Put coats and pantaloons into linen cases, made for the

332

purpose, and furnished with strings. Fill all crevices with small articles; as, if a trunk is not full, nor tightly packed, its contents will be shaken about, and get injured. A thin box, the exact size of the trunk, with a lid, and covered with brown linen, is a great convenience, to set inside, on the top of the trunk, to contain light articles which would be injured by tight packing. Have straps, with buckles, fastened to the inside, near the bottom, long enough to come up and buckle over this box. By this means, when a trunk is not quite full, this box can be strapped over so tight, as to keep the articles from rubbing. Under-clothing packs closer, by being rolled tightly, instead of being folded.

Bonnet-boxes, made of light wood, with a lock and key, are better than the paper bandboxes so annoying to travellers. Carpet bags are very useful, to carry the articles to be used on a journey. The best ones have sides inserted, iron rims, and a lock and key. A large silk travelling-bag, with a double linen lining, in which are stitched receptacles for toothbrush, combs, and other small articles, is a very convenient article for use when travelling.

A bonnet-cover, made of some thin material, like a large hood with a cape, is useful to draw over the bonnet and neck, to keep off dust, sun, and sparks from a steam engine. Green veils are very apt to stain bonnets, when damp.

In packing household furniture, for moving, have each box numbered, and then have a book, in which, as each box is packed, note down the number of the box, and the order in which its contents are packed, as this will save much labor and perplexity when unpacking. In packing china and glass, wrap each article,[Pg 317] separately, in paper, and put soft hay or straw at bottom and all around each. Put the heaviest articles at the bottom; and on the top of the box, write, "This side up."

CHAPTER XXXII.
ON THE CARE OF THE KITCHEN, CELLAR, AND STORE-ROOM.

If parents wish their daughters to grow up with good domestic habits, they should have, as one means of securing this result, a neat and cheerful kitchen. A kitchen should always, if possible, be entirely above ground, and well lighted. It should have a large sink, with a drain running under ground, so that all the premises may be kept sweet and clean. If flowers and shrubs be cultivated, around the doors and windows, and the yard near them be kept well turfed, it will add very much to their agreeable appearance. The walls should often be cleaned and whitewashed, to promote a neat look and pure air. The floor of a kitchen should be painted, or, which is better, covered with an oilcloth. To procure a kitchen oilcloth as cheaply as possible, buy cheap tow cloth, and fit it to the size and shape of the kitchen. Then have it stretched, and nailed to the south side of the barn, and, with a brush, cover it with a coat of thin rye paste. When this is dry, put on a coat of yellow paint, and let it dry for a fortnight. It is safest to first try the paint, and see if it dries well, as some paint never will dry. Then put on a second coat, and at the end of another fortnight, a third coat. Then let it hang two months, and it will last, uninjured, for many years. The longer the paint is left to dry, the better. If varnished, it will last much longer.

A sink should be scalded out every day, and occasionally with hot ley. On nails, over the sink, should be hung three good dish-cloths, hemmed, and furnished with loops; one for dishes not greasy, one for greasy[Pg 318] dishes, and one for washing pots and kettles. These should be put in the wash every week. The lady who insists upon this, will not be annoyed by having her dishes washed with dark, musty, and greasy, rags, as is too frequently the case.

Under the sink should be kept a slop-pail; and, on a shelf by it, a soap-dish and two water-pails. A large boiler, of warm soft water, should always be kept over the fire, well covered, and a hearth-broom and bellows be hung near the fire. A clock is a very important article in the kitchen, in order to secure regularity at meals.

No item of domestic labor is so frequently done in a negligent manner, by domestics, as this. A full supply of conveniences, will do much toward a remedy of this evil. A swab, made of strips of linen, tied to a stick, is useful to wash nice dishes, especially small, deep articles. Two or three towels, and three dish-cloths, should be used. Two large tin tubs, painted on the outside, should be provided; one for washing, and one for rinsing; also, a large old waiter, on which to drain the dishes. A soap-dish, with hard soap, and a fork, with which to use it, a slop-pail, and two pails for water, should also be furnished. Then, if there be danger of neglect, the following rules for washing dishes, legibly written, may be hung up by the sink, and it will aid in promoting the desired care and neatness.

Rules for Washing Dishes.

1. Scrape the dishes, putting away any food which may remain on them, and which it may be proper to save for future use. Put grease into the grease-pot, and whatever else may be on the plates, into the slop-pail. Save tea-leaves, for sweeping. Set all the dishes, when scraped, in regular piles; the smallest at the top.

2. Put the nicest articles in the wash-dish, and wash them in hot suds, with the swab or nicest dish-cloth. Wipe all metal articles, as soon as they are washed.[Pg 319] Put all the rest into the rinsing-dish, which should be filled with hot water. When they are taken out, lay them to drain on the waiter. Then rinse the dish-cloth, and hang it up, wipe the articles washed, and put them in their places.

3. Pour in more hot water, wash the greasy dishes with the dish-cloth made for them; rinse them, and set them to drain. Wipe them, and set them away. Wash the knives and forks, *being careful that the handles are never put in water*; wipe them, and then lay them in a knife-dish, to be scoured.

4. Take a fresh supply of clean suds, in which, wash the milk-pans, buckets, and tins. Then rinse and hang up this dish-cloth, and take the other; with which, wash the roaster, gridiron, pots, and kettles. Then wash and rinse the dish-cloth, and hang it up. Empty the slop-bucket and scald it. Dry metal teapots and tins before the

fire. Then put the fireplace in order, and sweep and dust the kitch-
en.

Some persons keep a deep and narrow vessel, in which to wash
knives with a swab, so that a careless domestic *cannot* lay them in
the water while washing them. This article can be carried into the
eating-room, to receive the knives and forks, when they are taken
from the table.

Kitchen Furniture.

Crockery. Brown earthen pans are said to be best, for milk and for
cooking. Tin pans are lighter, and more convenient, but are too cold
for many purposes. Tall earthen jars, with covers, are good to hold
butter, salt, lard, &c. Acids should never be put into the red earthen
ware, as there is a poisonous ingredient in the glazing, which the
acid takes off. Stone ware is better, and stronger, and safer, every
way, than any other kind.

Iron Ware. Many kitchens are very imperfectly supplied with the
requisite conveniences for cooking. When a person has sufficient
means, the following articles are all desirable. A nest of iron pots, of
[Pg 320]different sizes, (they should be slowly heated, when new;) a
long iron fork, to take out articles from boiling water; an iron hook,
with a handle, to lift pots from the crane; a large and small gridiron,
with grooved bars, and a trench to catch the grease; a Dutch oven,
called, also, a bakepan; two skillets, of different sizes, and a spider,
or flat skillet, for frying; a griddle, a waffle-iron, tin and iron bake
and bread-pans; two ladles, of different sizes; a skimmer; iron
skewers; a toasting-iron; two teakettles, one small and one large
one; two brass kettles, of different sizes, for soap-boiling, &c. Iron
kettles, lined with porcelain, are better for preserves. The German
are the best. Too hot a fire will crack them, but with care in this
respect, they will last for many years.

Portable furnaces, of iron or clay, are very useful, in Summer, in
washing, ironing, and stewing, or making preserves. If used in the
house, a strong draught must be made, to prevent the deleterious
effects of the charcoal. A box and mill, for spice, pepper, and coffee,
are needful to those who use these articles. Strong knives and forks,

a sharp carving-knife, an iron cleaver and board, a fine saw, steel-yards, chopping-tray and knife, an apple-parer, steel for sharpening knives, sugar-nippers, a dozen iron spoons, also a large iron one with a long handle, six or eight flatirons, one of them very small, two iron-stands, a ruffle-iron, a crimping-iron, are also desirable.

Tin Ware. Bread-pans, large and small pattypans, cake-pans, with a centre tube to insure their baking well, pie-dishes, (of block-tin,) a covered butter-kettle, covered kettles to hold berries, two sauce-pans, a large oil-can, (with a cock,) a lamp-filler, a lantern, broad-bottomed candlesticks for the kitchen, a candle-box, a funnel or tunnel, a reflector, for baking warm cakes, an oven or tin-kitchen, an apple-corer, an apple-roaster, an egg-boiler, two sugar-scoops, and flour and meal-scoop, a set of mugs, three dippers, a pint, quart, and gallon measure, a set of scales and weights, three or four pails, painted on the outside, a slop-bucket, with a[Pg 321] tight cover, painted on the outside, a milk-strainer, a gravy-strainer, a colander, a dredging-box, a pepper-box, a large and small grater, a box, in which to keep cheese, also a large one for cake, and a still larger one for bread, with tight covers. Bread, cake, and cheese, shut up in this way, will not grow dry as in the open air.

Wooden Ware. A nest of tubs, a set of pails and bowls, a large and small sieve, a beetle for mashing potatoes, a spad or stick for stirring butter and sugar, a bread-board, for moulding bread and making pie-crust, a coffee-stick, a clothes-stick, a mush-stick, a meat-beetle to pound tough meat, an egg-beater, a ladle for working butter, a bread-trough, (for a large family,) flour-buckets, with lids to hold sifted flour and Indian meal, salt-boxes, sugar-boxes, starch and indigo-boxes, spice-boxes, a bosom-board, a skirt-board, a large ironing-board, two or three clothes-frames, and six dozen clothes-pins.

Basket Ware. Baskets, of all sizes, for eggs, fruit, marketing, clothes, &c.; also chip-baskets. When often used, they should be washed in hot suds.

Other Articles. Every kitchen needs a box containing balls of brown thread and twine, a large and small darning needle, rolls of waste-paper and old linen and cotton, and a supply of common holders. There should also be another box, containing a hammer, carpet-tacks, and nails of all sizes, a carpet-claw, screws and a screw-driver, pincers, gimlets of several sizes, a bed-screw, a small saw, two chisels, (one to use for buttonholes in broadcloth,) two awls, and two files.

In a drawer, or cupboard, should be placed, cotton table-cloths, for kitchen use, nice crash towels, for tumblers, marked, T T; coarser towels, for dishes, marked, T; six large roller-towels; a dozen hand-towels, marked, H T; and a dozen hemmed dish-cloths, with loops. Also, two thick linen pudding or dumpling-cloths, a gelly-bag, made of white flannel, to strain gelly, a starch-strainer, and a bag for boiling clothes.

[Pg 322]

In a closet, should be kept, arranged in order, the following articles: the dust-pan, dust-brush, and dusting-cloths, old flannel and cotton for scouring and rubbing, sponges for washing windows and looking-glasses, a long brush for cobwebs, and another for washing the outside of windows, whisk-brooms, common brooms, a coat-broom or brush, a whitewash-brush, a stove-brush, shoebrushes and blacking, articles for cleaning tin and silver, leather for cleaning metals, bottles containing stain-mixtures, and other articles used in cleansing.

ON THE CARE OF THE CELLAR.

A cellar should often be whitewashed, to keep it sweet. It should have a drain, to keep it perfectly dry, as standing water, in a cellar, is a sure cause of disease in a family. It is very dangerous to leave decayed vegetables in a cellar. Many a fever has been caused, by the poisonous miasm thus generated. The following articles are desirable in a cellar: a safe, or moveable closet, with sides of wire or perforated tin, in which cold meats, cream, and other articles should be kept; (if ants be troublesome, set the legs in tin cups of water;) a refrigerator, or large wooden box, on feet, with a lining of tin or zinc, and a space between the tin and wood filled with powdered charcoal, having at the bottom, a place for ice, a drain to carry off the water, and also moveable shelves and partitions. In this, articles are kept cool. It should be cleaned, once a week. Filtering jars, to purify water, should also be kept in the cellar. Fish and cabbages, in a cellar, are apt to scent a house, and give a bad taste to other articles.

STOREROOM.

Every house needs a storeroom, in which to keep tea, coffee, sugar, rice, candles, &c. It should be furnished with jars, having labels, a large spoon, a fork, sugar and flour-scoops, a towel, and a dish-cloth.

[Pg 323]

Modes of destroying Insects and Vermin.

Bed-bugs should be kept away, by filling every chink in the bedstead with putty, and, if it be old, painting it over. Of all the mixtures for killing them, *corrosive sublimate and alcohol* is the surest. This is a strong poison.

Cockroaches may be destroyed, by pouring boiling water into their haunts, or setting a mixture of arsenic, mixed with Indian meal and molasses, where they are found. Chloride of lime and sweetened water will also poison them.

Fleas. If a dog be infested with these insects, put him in a tub of warm soapsuds, and they will rise to the surface. Take them off, and burn them. Strong perfumes, about the person, diminish their attacks. When caught between the fingers, plunge them in water, or they will escape.

Crickets. Scalding, and sprinkling Scotch snuff about the haunts of these insects, are remedies for the annoyance caused by them.

Flies can be killed, in great quantities, by placing about the house vessels, filled with sweetened water and *cobalt.* Six cents worth of cobalt is enough for a pint of water. It is very poisonous.

Musquitoes. Close nets around a bed, are the only sure protection at night, against these insects. Spirit of hartshorn is the best antidote for their bite. Salt and water is good.

Red or *Black Ants* may be driven away, by scalding their haunts, and putting Scotch snuff wherever they go for food. Set the legs of closets and safes in pans of water, and they cannot get at them.

Moths. Airing clothes does not destroy moths, but laying them in a hot sun does. If articles be tightly sewed up in linen, and fine tobacco be put about them, it is a sure protection. This should be done in April.

Rats and Mice. A good cat is the best remedy for[Pg 324] these annoyances. Equal quantities of hemlock, (or *cicuta,*) and old cheese, will poison them, but this renders the house liable to the inconvenience of a bad smell. This evil, however, may be lessened, by placing a dish, containing oil of vitriol poured on saltpetre, where the smell is most annoying. Chloride of lime and water is also good.

CHAPTER XXXIII.
ON SEWING, CUTTING, AND MENDING.

Every young girl should be taught to do the following kinds of stitch, with propriety. Over-stitch, hemming, running, felling, stitching, back-stitch and run, buttonhole-stitch, chain-stitch, whipping, darning, gathering, and cross-stitch.

In doing over-stitch, the edges should always be first fitted, either with pins or basting, to prevent puckering. In turning wide hems, a paper measure should be used, to make them even. Tucks, also, should be regulated by a paper measure. A fell should be turned, before the edges are put together, and the seam should be over-sewed, before felling. All biased or goring seams should be felled. For stitching, draw a thread, and take up two or three threads at a stitch.

In making buttonholes, it is best to have a pair of scissors, made for the purpose, which cut very neatly. For broadcloth, a chisel and board are better. The best stitch is made by putting in the needle, and then turning the thread around it, near the eye. This is better than to draw the needle through, and then take up a loop. A thread should first be put across each side of the buttonhole, and also a stay-thread, or bar, at each end, before working it. In working the buttonhole, keep the stay-thread as far from the edge as possible. A small bar should be worked at each end.[Pg 325] Whipping is done better by sewing *over*, and not under. The roll should be as fine as possible, the stitches short, the thread strong, and in sewing, every gather should be taken up.

The rule for *gathering*, in shirts, is, to draw a thread, and then take up two threads and skip four. In *darning*, after the perpendicular threads are run, the crossing threads should interlace, exactly, taking one thread and leaving one, like woven threads.

The neatest sewers always fit and baste their work, before sewing; and they say they always save time in the end, by so doing, as they never have to pick out work, on account of mistakes.

It is wise to sew closely and tightly all new garments, which will never be altered in shape; but some are more nice than wise, in sewing frocks, and old garments, in the same style. However, this is the least common extreme. It is much more frequently the case, that articles, which ought to be strongly and neatly made, are sewed so that a nice sewer would rather pick out the threads and sew over again, than to be annoyed with the sight of grinning stitches, and vexed with constant rips.

Workbaskets. It is very important to neatness, comfort, and success in sewing, that a lady's workbasket should be properly fitted up. The following articles are needful to the mistress of a family: a large basket, to hold work; having in it, fastened, a smaller basket, or box, containing a needle-book, in which are needles of every size, both blunts and sharps, with a larger number of those sizes most used; also, small and large darning-needles, for woollen, cotton, and silk; two tape-needles, large and small; nice scissors, for fine work; buttonhole scissors; an emery-bag; two balls of white and yellow wax; and two thimbles, in case one should be mislaid. When a person is troubled with damp fingers, a lump of soft chalk, in a paper, is useful, to rub on the ends of the fingers.

Besides this box, keep in the basket, common scissors; [Pg 326]small shears; a bag containing tapes, of all colors and sizes, done up in rolls; bags, one, containing spools of white, and another of colored, cotton thread, and another for silks, wound on spools or papers; a box or bag for nice buttons, and another for more common ones; a bag containing silk braid, welting cords, and galloon binding. Small rolls of pieces of white and brown linen and cotton, are also often needed. A brick pincushion is a great convenience, in sewing, and better than screw-cushions. It is made by covering half a brick with cloth, putting a cushion on the top, and covering it tastefully. It is very useful to hold pins and needles, while sewing, and to fasten long seams when basting and sewing.

To make a Frock. The best way for a novice, is, to get a dress fitted (not sewed) at the best mantuamaker's. Then take out a sleeve, rip it to pieces, and cut out a pattern. Then take out half of the waist, (it must have a seam in front,) and cut out a pattern of the back and fore-body, both lining and outer part. In cutting the patterns, iron the pieces, smooth, let the paper be stiff, and, with a pin, prick holes in the paper, to show the gore in front, and the depth of the seams.

With a pen and ink, draw lines from each pinhole, to preserve this mark. Then baste the parts together again, in doing which, the un-basted half will serve as a pattern. When this is done, a lady of common ingenuity can cut and fit a dress, by these patterns. If the waist of a dress be too tight, the seam under the arm must be let out; and in cutting a dress, an allowance should be made, for letting it out, if needful, at this seam. The lining of the fore-body must be biased.

The linings for the waists of dresses should be stiffened cotton or linen. In cutting bias-pieces, for trimming, they will not set well, unless they are exact. In cutting them, use a long rule, and a lead pencil or piece of chalk. Welting-cords should be covered with bias-pieces; and it saves time, in many cases, to baste[Pg 327] on the welting-cord, at the same time that you cover it. The best way to put on hooks and eyes, is to sew them on double broad tape, and then sew this on the frock-lining. They can then be moved easily, and do not show where they are sewed on.

In cutting a sleeve, double it biased. The skirts of dresses look badly, if not full; and in putting on lining, at the bottom, be careful to have it a very little fuller than the dress, or it will shrink, and look badly. All thin silks look much better with lining, and last much longer, as do aprons, also. In putting a lining to a dress, baste it on each separate breadth, and sew it in at the seams, and it looks much better than to have it fastened only at the bottom. Make notches in selvedge, to prevent it from drawing up the breadth. Dresses, which are to be washed, should not be lined.

Figured silks do not generally wear well, if the figure be large and satin-like. Black and plain-colored silks can be tested, by procuring samples, and making creases in them; fold the creases in a bunch, and rub them against a rough surface, of moreen or carpeting. Those which are poor, will soon wear off, at the creases. Plaids look becoming, for tall women, as they shorten the appearance of the figure. Stripes look becoming, on a large person, as they reduce the apparent size. Pale persons should not wear blue or green, and bru-nettes should not wear light delicate colors, except shades of buff, fawn, or straw color. Pearl white is not good for any complexion. Dead white and black look becoming on almost all persons. It is best to try colors, by candle-light, for evening dresses; as some colors, which look very handsome in the daylight, are very homely when seen by candle-light. Never cut a dress low in the neck, as this shows that a woman is not properly instructed in the rules of modesty and decorum, or that she has not sense enough to regard them. Never be in haste to be first in a fashion, and never go to the extremes.

[Pg 328]

In buying linen, seek for that which has a round close thread, and is perfectly white; for, if it be not white, at first, it will never afterwards become so. Much that is called linen, at the shops, is half cotton, and does not wear so well as cotton alone. Cheap linens are usually of this kind. It is difficult to discover which are all linen; but the best way, is, to find a lot, presumed to be good, take a sample, wash it, and ravel it. If this be good, the rest of the same lot will probably be so. If you cannot do this, draw a thread, each way, and if both appear equally strong, it is probably all linen. Linen and cotton must be put in clean water, and boiled, to get out the starch, and then ironed. A long piece of linen, a yard wide, will, with care and calculation, make eight shirts. In cutting it, take a shirt of the right size, as a guide, in fitting and basting. Bosom-pieces, false collars, &c. must be cut and fitted, by a pattern which suits the person for whom the articles are designed. Gentlemen's night-shirts are made like other shirts, except that they are longer. In cutting chemises, if the cotton or linen is a yard wide, cut off small half gores, at the top of the breadths, and set them on the bottom. Use a long rule and a pencil, in cutting gores. In cutting cotton, which is quite wide, a seam can be saved, by cutting out two at once, in this manner:—cut off three breadths, and, with a long rule and a pencil, mark and cut off the gores, thus: from one breadth, cut off two gores, the whole length, each gore one fourth of the breadth, at the bottom, and tapering off to a point, at the top. The other two breadths are to have a gore cut off from each, which is one fourth wide at top, and two fourths at bottom. Arrange these pieces right, and they will make two chemises, one having four seams, and the other three. This is a much easier way of cutting, than sewing the three breadths together, in bag-fashion, as is often done. The biased, or goring seams, must always be felled. The sleeves and neck can be

346

cut according to the taste of the wearer, by another chemise[Pg 329] for a pattern. There should be a lining around the armholes, and stays at all corners. Six yards, of yard width, will make two chemises.

Old silk dresses, quilted for skirts, are very serviceable. White flannel is soiled so easily, and shrinks so much in washing, that it is a good plan to color it a light dove-color, according to the receipt given on page 301. Cotton flannel, dyed thus, is also good for common skirts. In making up flannel, back-stitch and run the seams, and then cross-stitch them open. Nice flannel, for infants, can be ornamented, with very little expense of time, by turning up the hem, on the right side, and making a little vine at the edge, with saddler's silk. The stitch of the vine is a modification of buttonhole-stitch.

Long night gowns are best, cut a little goring. It requires five yards, for a long nightgown, and two and a half for a short one. Linen nightcaps wear longer than cotton ones, and do not, like them, turn yellow. They should be ruffled with linen, as cotton borders will not last so long as the cap. A double-quilted wrapper is a great comfort, in case of sickness. It may be made of two old dresses. It should not be cut full, but rather like a gentleman's study-gown, having no gathers or plaits, but large enough to slip off and on with ease. A double gown, of calico, is also very useful. Most articles of dress, for grown persons or children, require patterns.

Bedding. The best beds, are thick hair mattresses, which, for persons in health, are good for Winter as well as Summer use. Mattresses may also be made of husks, dried and drawn into shreds; also, of alternate layers of cotton and moss. The most profitable sheeting, is the Russian, which will last three times as long as any other. It is never perfectly white. Unbleached cotton is good for Winter. It is poor economy to make narrow and short sheets, as children and domestics will always slip them off, and soil the bed-tick and bolster. They should be three yards long, and two[Pg 330] and a half wide, so that they can be tucked in all around. All bed-linen should be marked and numbered, so that a bed can always be made properly, and all missing articles be known.

Mending. Silk dresses will last much longer, by ripping out the sleeves, when thin, and changing the arms, and also the breadths of the skirt. Tumbled black silk, which is old and rusty, should be dipped in water, then be drained for a few minutes, without squeezing or pressing, and then ironed. Cold tea is better than water. Sheets, when worn thin in the middle, should be ripped, and the other edges sewed together. Window-curtains last much longer, if lined, as the sun fades and rots them. Broadcloth should be cut with reference to the way the nap runs. When pantaloons are thin, it is best to newly seat them, cutting the piece inserted in a curve, as corners are difficult to fit. When the knees are thin, it is a case of domestic surgery, which demands *amputation.* This is performed, by cutting off both legs, some distance above the knees, and then changing the legs. Take care to cut them off exactly of the same length, or in the exchange they will not fit. This method brings the worn spot under the knees, and the seam looks much better than a patch and darn. Hose can be cut down, when the feet are worn. Take an old stocking, and cut it up for a pattern. Make the heel short. In sewing, turn each edge, and run it down, and then sew over the edges. This is better than to stitch and then cross-stitch. Run thin places in stockings, and it will save darning a hole. If shoes are worn through on the sides, in the upper-leather, slip pieces of broadcloth under, and sew them around the holes. If, in sewing, the thread kinks, break it off and begin at the other end. In using spool-cotton, thread the needle with the end which comes off first, and not the end where you break it off. This often prevents kinks.

[Pg 331]

CHAPTER XXXIV.
ON THE CARE OF YARDS AND GARDENS.

The authorities consulted in the preparation of this and kindred chapters, are, Loudon's Encyclopædia of Gardening, Bridgeman's Young Gardener, Hovey's Magazine of Horticulture, the writings of Judge Buel,[T] and Downing's Landscape Gardening.

On the Preparation of Soil.

If the garden soil be clayey, and adhesive, put on a covering of sand, three inches thick, and the same depth of well-rotted manure. Spade it in, as deep as possible, and mix it well. If the soil be sandy and loose, spade in clay and ashes. Ashes are good for all kinds of soil, as they loosen those which are close, hold moisture in those which are sandy, and destroy insects. The best kind of soil, is that, which will hold water the longest, without becoming hard, when dry.

To prepare Soil for Pot-plants, take one fourth part of common soil, one fourth part of well-decayed manure, and one half of vegetable mould, from the woods, or from a chip-yard. Break up the manure, fine, and sift it through a lime-screen, (or coarse wire sieve.) These materials must be thoroughly mixed. When the common soil which is used, is adhesive, and, indeed, in most other cases, it is necessary to add sand, the proportion of which, must depend on the nature of the soil.

On the Preparation of a Hot Bed. Dig a pit, six feet long, five feet wide, and thirty inches deep. Make a frame, of the same size, with the back two feet high, the front fifteen inches, and the sides sloped from the back to the front. Make two sashes, each three feet [Pg 332]by five, with the panes of glass lapping like shingles, instead of having cross bars. Set the frame over the pit, which should then be filled with fresh horse-dung, which has not lain long, nor been sodden by water. Tread it down, hard, then put into the frame, light, and very rich soil, ten or twelve inches deep, and cover it with the sashes, for two or three days. Then stir the soil, and sow the seeds in shallow drills, placing sticks by them, to mark the different kinds.

Keep the frame covered with the glass, whenever it is cold enough to chill the plants; but at all other times, admit fresh air, which is indispensable to their health. When the sun is quite warm, raise the glasses, enough to admit air, and cover them with matting or blankets, or else the sun may kill the young plants. Water the bed at evening, with water which has stood all day, or, if it be fresh drawn, add a little warm water. If there be too much heat in the bed, so as to scorch or wither the plants, make deep holes, with stakes, and fill them up when the heat is reduced. In very cold nights, cover the box with straw.

On Planting Flower Seeds.

Break up the soil, till it is very soft, and free from lumps. Rub that nearest the surface, between the hands, to make it fine. Make a circular drill, a foot in diameter. For seeds as large as sweet peas, it should be half an inch deep. The smallest seeds must be planted very near the surface, and a very little fine earth be sifted over them. Seeds are to be planted either deeper or nearer the surface, according to their size. After covering them with soil, beat them down with a trowel, so as to make the earth as compact as it is after a heavy shower. Set up a stick, in the middle of the circle, with the name of the plant heavily written upon it, with a dark lead pencil. This remains more permanent, if white lead be first rubbed over the surface. Never plant, when the soil is very wet. In very dry times, water the seeds at night. Never use[Pg 333] very cold water. When the seeds are small, many should be planted together, that they may assist each other in breaking the soil. When the plants are an inch high, thin them out, leaving only one or two, if the plant be a large one, like the Balsam; five or six, when it is of a medium size; and eighteen or twenty of the smaller size. Transplanting, retards the growth of a plant about a fortnight. It is best to plant at two different times, lest the first planting should fail, owing to wet or cold weather.

To Plant Garden Seeds.

Make the beds a yard wide; lay across them a board, a yard long and a foot wide, and, with a stick, make a furrow, on each side of it, one inch deep. Scatter the seeds in this furrow, and cover them. Then lay the board over them and step on it, to press down the earth. When the plants are an inch high, thin them out, leaving spaces proportioned to their sizes. Seeds of a similar species, such as melons and squashes, should not be planted very near to each other, as this causes them to degenerate. The same kinds of vegetables should not be planted in the same place, for two years in succession.

On Transplanting.

Transplant at evening, or, which is better, just before a shower. Take a round stick, sharpened at the point, and make openings to receive the plants. Set them a very little deeper than they were before, and press the soil firmly round them. Then water them, and cover them for three or four days, taking care that sufficient air be admitted. If the plant can be removed, without disturbing the soil around the root, it will not be at all retarded, by transplanting. Never remove leaves and branches, unless a part of the roots be lost.

To Re-pot House-Plants.

Renew the soil, every year, soon after the time[Pg 334] of blossoming. Prepare soil, as previously directed. Loosen the earth from the pot, by passing a knife around the sides. Turn the plant upside down, and remove the pot. Then remove all the matted fibres at the bottom, and all the earth, except that which adheres to the roots. From woody plants, like roses, shake off all the earth. Take the new pot, and put a piece of broken earthen-ware over the hole at the bottom; and then, holding the plant in the proper position, shake in the earth, around it. Then pour in water, to settle the earth, and heap on fresh soil, till the pot is even full. Small pots are considered better than large ones, as the roots are not so likely to rot, from excess of moisture.

On the Laying out of Yards and Gardens.

In planting trees, in a yard, they should be arranged in groups, and never planted in straight lines, nor sprinkled about, as solitary trees. The object of this arrangement, is, to imitate Nature, and secure some spots of dense shade and some of cleared turf. In yards which are covered with turf, beds can be cut out of it, and raised for flowers. A trench should be made around, to prevent the grass from running on them. These beds can be made in the shape of crescents, ovals, or other fanciful forms, of which, the figure below is one specimen.

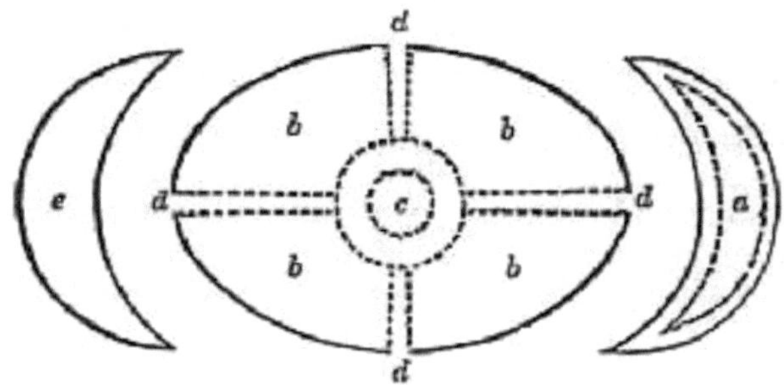

Fig. 39.

In laying out beds, in gardens and yards, a very pretty bordering can be made, by planting them with common flax seed, in a line about three inches from the edge. This can be trimmed, with shears, when it grows too high.

[Pg 335]

On the Cultivation of Bulbs, and Tuberous Roots.

For planting the *Amaryllis*, take one third part of leaf mould, half as much sand, and the remainder, earth from under fresh grass sods. Plant them in May. The bulb should not be set more than half its depth in the ground.

The *Anemone* and *Ranunculus* are medium, or half-hardy, roots. They should be planted in soil which is enriched with cowdung, and the beds should be raised only an inch from the walk. They must be planted in October, in drills, two inches deep, the claws of the roots downward, and be shaded when they begin to bud.

The *Crocus* must be planted in October, two inches deep, and four inches apart. In measuring the depth, always calculate from the top of the bulb.

Crown Imperial. This must be planted in September, three or four inches deep; and need not be taken up but once in three years.

Gladiolus. Those who have greenhouses, or pits, plant the Gladiolus in October, and preserve it in pots through the Winter. Those who have not these conveniences, may plant these bulbs late in April. The earth must be composed of one half common soil, one fourth leaf mould, and one fourth sand. Plant them about an inch deep.

Hyacinths should be planted in October, eight inches apart, and three or four inches deep, in a rich soil.

Jonquilles should be planted in October, two inches deep, in a rich soil, and should not be taken up oftener than once in three years.

Narcissus. This should be planted in October, four inches deep; covered, through the Winter, with straw and leaves, six inches thick; and uncovered in the middle of March.

Oxalis. Plant this in September, in a soil, composed of two thirds common earth, and one third leaf mould. The old bulb dies after blossoming, and is succeeded by a new one.

[Pg 336]

Plant *Tulips*, in rich soil, in October, three inches deep.

Plant *Tuberoses* late in April, in a rich, sandy soil. They are delicate plants, and should be covered, in case of frosts.

Daffodils should be planted two inches deep.

When bulbs have done flowering, and their leaves begin to decay, they should be taken up and dried, and kept in a dry place, till October, when they are to be replanted, taking off the offsets, and putting them in a bed by themselves.

Bulbs which blossom in water, or are in any other way forced to bloom out of season, are so much exhausted by it, that it takes them two or three years to recover their beauty.

Dahlias. Dig a hole, a foot and a half deep; fill it with very light, loose, and rich, soil; and drive in a stake, a yard and a half high, to which, to tie the future plants. Then set in the root, so that it shall be an inch below the soil, where the sprout starts. When the plants are two feet high, tie them to the stakes, and take off some of the lower side-shoots. Continue to tie them, as their growth advances. If the roots are planted in the open borders, without any previous growth, it should be done as early as the first of May, and they should be covered from the frosts. When they are brought forward, in pots or hot-beds, they should be put out, in the middle of June. It is said, by gardeners, that late planting, is better than early, for producing perfect flowers. In the Autumn, after the frosts have destroyed the tops, let the roots remain awhile in the ground, to ripen; then dig them up, and pack them away, in some place where they will neither mould, from dampness, nor freeze. In the Spring, these roots will throw out sprouts, and must then be divided, so as to leave a good shoot, attached to a piece of the tuber or old stem, and each shoot will make a new plant. It is stated, that if the shoots themselves, without any root, be planted in light soil, covered with a[Pg 337] bell-glass, or large tumbler, and carefully watered, they will produce plants superior to those with roots.

Annuals

These are flowers which last only one season. They should be so planted, that the tallest may be in the middle of a bed, and the shortest at the edges; and flowers of a similar color should not be planted adjacent to each other.

The following is a list of some of the handsomest Annuals, arranged with reference to their color and height. Those with a star before them, do best when sowed in the Autumn. Those with *tr.* after them, are trailing plants.

SIX INCHES TO ONE FOOT HIGH.

White. Ice Plant, Sweet Alyssum, White Leptosiphon, Walker's Schizopetalon, Blumenbachia insignis, *Candytuft.

Yellow. *Yellow Chryseis or Eschscholtzia, Sanvitalia procumbens, *tr.*, Musk-flowered Mimulus.

Rose. Many-flowered Catchfly, Rose-colored Verbena, *tr.*

Red. *Chinese Annual Pink, Virginian Stock, Calandrinia Speciosa.

Blue. Graceful Lobelia, Nemophila insignis, Clintonia pulchella, Clintonia elegans, Nolana atriplicifolia, *tr.*, Anagallis indica, Commelina cœlestis, Grove Love, Pimpernel (blue.)

Varying Colors. *Heart's Ease, or Pansy, Dwarf Love in a Mist, *Rose Campion.

ONE FOOT TO EIGHTEEN INCHES HIGH.

White. Venus's Looking Glass, Priest's Schizanthus, Sweet-scented Stevia, White Evening Primrose.

Yellow. Drummond's Coreopsis, *New Dark Coreopsis, Golden Hawkweed, Dracopis amplexicaulis, Drummond's Primrose, Cladanthus arabicus, Peroffsky's Erysimum.

[Pg 338]

Rose. Drummond's Phlox, Rodanthe, Rose-colored Nonea, Clarkia rosea, Silene Tenorei, Silene armeria.

Red. Crimson Coxcomb, Silene pendula, Crimson Dew Plant, *tr.*

Scarlet. Cacalia coccinea, Flos Adonis, Scarlet Zinnia, Mexican Cuphea.

Lilac and Purple. Clarkia elegans, Clarkia pulchella, *Purple Candytuft, *Purple Petunia, *tr.*, *Crimson Candytuft, Double Purple Jacobæa, Leptosiphon androsaceus, all the varieties of Schizanthus, Veined Verbena, *tr.*, *Purple eternal Flower.

Blue. Ageratum Mexicanum, *Gilia capitata, Spanish Nigella, Blue Eutoca, Dwarf Convolvulus, Didiscus cœruleus.

Lilac, Purple, or *Blue and White.* Collinsia bicolor, Gilia tricolor.

Very Dark. Lotus Jacobæus, Salpiglossis, Scabious.

Colors varying. German Aster, Balsam, Rocket Larkspur, Ten-week Stock, Poppy.

EIGHTEEN INCHES TO TWO FEET.

White. *White Petunia, *tr.*, White Clarkia, Double White Jacobæa, Love in a Mist.

Red. *Lavatera trimestris, Red Zinnia, Malva miniata.

Lilac and Purple. Globe Amaranthus, Purple Sweet Sultan, Sweet Scabious, Purple Zinnia, Prince's Feather, Large Blue Lupine, *Catchfly.

TWO FEET AND UPWARDS.

White. Winged Ammobium, *White Lavatera, White Sweet Sultan, *New White Eternal Flower, White Helicrysum, *White Larkspur.

Yellow. Golden Bartonia, *Golden Coreopsis, Yellow Sweet Sultan, African Marigold, Yellow Argemone, French Marigold, Yellow Coxcomb, Yellow Hibiscus.

The Malope grandiflora and the Cleome are fine tall annuals.

[Pg 339]

Climbing Plants.

The following are the most beautiful *annual climbers*: Crimson, and White, Cypress Vine; White, and Buff, Thunbergia; Scarlet Flowering Bean; Hyacinth Bean Loasa; Morning Glory; Crimson, and Spotted, Nasturtium; Balloon Vine; Sweet Pea; Tangier Pea; Lord Anson's Pea; Climbing Cobæa; Pink, and White, Maurandia.

The following are the most valuable *perennial climbers*: Sweet-scented Monthly Honeysuckle; Yellow, White, and Coral, Honeysuckles; Purple Glycine; Clematis; Bitter Sweet; Trumpet Creeper.

The Everlasting Pea is a beautiful perennial climber. The Climbing Cobæa, and Passion Flower, are also beautiful perennials, but must be protected in Winter.

356

Perennials.

Those who cannot afford every year to devote the time necessary to the raising of annuals, will do well to supply their borders with perennials. The following is a list of some of those generally preferred.

Adonis, yellow; Columbine, all colors; Alyssum, yellow; Asclepias, orange and purple; Bee Larkspur, blue; Perennial Larkspur, all colors; Cardinal Flower, scarlet; Chinese Pink, various colors; Clove Pink; Foxglove, purple and white; Gentian, purple and yellow; Hollyhock, various colors; *Lily of the Valley; American Phlox, various colors; Scarlet Lychnis; Monkshood, white and blue; *Spirea, white, and pink; *Ragged Robin, pink; Rudbeckia, yellow, and purple; Sweet William, in variety. Those marked with a star cannot be obtained from seed, but must be propagated by roots, layers, &c.

Herbaceous Roots.

These are such as die to the root, in the Fall, and come up again in the Spring, such as Pæonies, crimson, white, sweet-scented, and straw-colored; Artemisia,[Pg 340] of many colors; White and Purple Fleur-de-lis; White, Tiger, Fire, and other Lilies; Little Blue Iris; Chrysanthemums, &c. These are propagated by dividing the roots.

Shrubs.

The following are the finest *Shrubs for yards*: Lilacs, (which, by budding, can have white and purple on the same tree,) Double Syringas, Double Althæas, Corchorus Japonicus, Snow-berry, Double-flowering Almond, Pyrus Japonica, Common Barberry, Burning Bush, Rose Acacia, Yellow Laburnum. The following are the finest Roses: Moss Rose, White, and Red; Double and Single Yellow Rose, (the last needs a gravelly soil and northern exposure;) Yellow Multiflora; La Belle Africana; Small Eglantine, for borders; Champney's Blush Rose; Noisette; Greville, (very fine;) Damask; Blush, White, and Cabbage Roses. Moss Roses, when budded on other rose bushes, last only three years.

Shade Trees. The following are among the finest: Mountain Ash; Ailanthus, or Tree of Heaven, (grows very fast;) Tulip Tree; Linden; Elm; Locust; Maple; Dog Wood; Horse Chestnut; Catalpa; Hemlock; Silver Fir; and Cedar. These should be grouped, in such a manner that trees of different shades of green, and of different heights, should stand in the same group.

The Autumn is the best time for transplanting trees. Take as much of the root, as possible, especially the little fibres, which should never become dry. If kept long, before they are set out, put wet moss around them, and water them. Dig holes, larger than the extent of the roots; let one person hold the tree in its former position, and another place the roots, carefully, as they were before, cutting off any broken or wounded root. *Be careful not to let the tree be more than an inch deeper than it was before.* Let the soil be soft, and well manured; shake the tree, as the soil is shaken[Pg 341] in, that it may mix well among the small fibres. Do not tread the earth down, while filling the hole; but, when it is full, raise a slight mound, of, say, four inches, and then tread it down. Make a little basin, two inches deep, around the stem, to hold water, and fill it. Never cut off leaves nor branches, unless some of the roots are lost. Tie the trees to a stake, and they will be more likely to live. Water them often.

On the Care of House-Plants.

The soil of house-plants should be renewed every year, as previously directed. In Winter, they should be kept as dry as they can be without wilting. Many house-plants are injured by giving them too much water, when they have little light and fresh air. This makes them grow spindling. The more fresh air, warmth, and light, they have, the more water is needed. They ought not to be kept very warm in Winter, nor exposed to great changes of atmosphere. Forty degrees is a proper temperature for plants in Winter, when they have little sun and air. When plants have become spindling, cut off their heads, entirely, and cover the pot in the earth, where it has the morning sun, only. A new and flourishing head will spring out. Few house-plants can bear the sun at noon. When insects infest plants, set them in a closet, or under a barrel, and burn tobacco. The smoke kills any insect enveloped in it. When plants are frozen, cold water, and a gradual restoration of warmth, are the best remedies. Never use very cold water for plants, at any season.

FOOTNOTE:

[T] His 'Farmers' Companion' was written expressly for the larger series of 'The School Library,' issued by the publishers of this volume.

CHAPTER XXXV.
ON THE PROPAGATION OF PLANTS.

Bulbous roots are propagated by offsets; some growing on the top, others around the sides. Many[Pg 342] plants are propagated by cutting off twigs, and setting them in earth, so that two or three eyes are covered. To do this, select a side shoot, ten inches long, two inches of it, being of the preceding year's growth, and the rest, the growth of the season when it is set out. Do this, when the sap is running, and put a piece of crockery at the bottom of the shoot, when it is buried. One eye, at least, must be under the soil. Water it, and shade it in hot weather. Plants are also propagated by layers. To do this, take a shoot, which comes up near the root, bend it down, so as to bring several eyes under the soil, leaving the top above ground. If the shoot be cut half through, in a slanting direction, at one of these eyes, before burying it, the result is more certain. Roses, honeysuckles, and many other shrubs, are readily propagated thus. They will generally take root, by being simply buried; but cutting them, as here directed, is the best method. Layers are more certain than cuttings. For all woody plants, budding and grafting are favorite methods of propagation. In all such plants, there is an outer and inner bark; the latter containing the sap vessels, in which the nourishment of the tree ascends.

The success of grafting, or inoculating, consists in so placing the bud or graft, that the sap vessels of the inner bark shall exactly join those of the plant into which they are grafted, so that the sap may pass from one into the other.

The following are directions for *budding*, which may be performed at any time from July to September.

Select a smooth place, on the stock into which you are to insert the bud. Make a horizontal cut, across the rind, through to the firm wood; and from the middle of this, make a slit downward, perpendicularly, an inch or more long, through to the wood. Raise the bark of the stock, on each side of the perpendicular cut, for the admission of the bud, as is shown in the annexed engraving, (Fig. 40.) Then take a shoot of this year's growth, and slice from it a bud, taking an[Pg 343] inch below and an inch above it, and some portion of the wood under it. Then carefully slip off the woody part, under the bud. Examine whether the eye or gem of the bud be perfect. If a little hole appears in that part, the bud has lost its root, and another must be selected. Insert the bud, so that *a*, of the bud, shall pass to a, of the stock; then *b*, of the bud, must be cut off, to match the cut, b, in the stock, and fitted exactly to it, as it is this alone which insures success. Bind the parts, with fresh bass, or woollen yarn, beginning a little below the bottom of the perpendicular slit, and winding it closely round every part, except just over the eye of the bud, until you arrive above the horizontal cut. Do not bind it too tightly, but just sufficient to exclude air, sun, and wet. This is to be removed, after the bud is firmly fixed, and begins to grow.

Fig.
40.

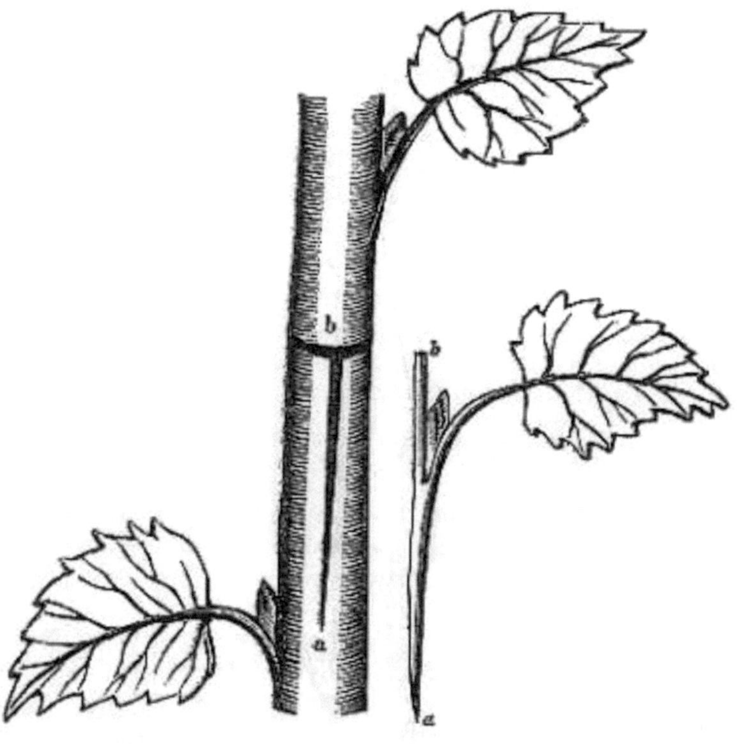

Seed-fruit can be budded into any other seed-fruit, and stone-fruit into any other stone-fruit; but stone and seed-fruits, cannot be thus mingled.

[Pg 344]

Rose bushes can have a variety of kinds budded into the same stock. Hardy roots are the best stocks. The branch above the bud, must be cut off, the next March or April after the bud is put in. Apples and pears, are more easily propagated by ingrafting, than by budding.

Ingrafting is a similar process to budding, with this advantage; that it can be performed on large trees, whereas budding can be applied only on small ones. The two common kinds of ingrafting, are whip-grafting, and split-grafting. The first kind is for young trees, and the other for large ones.

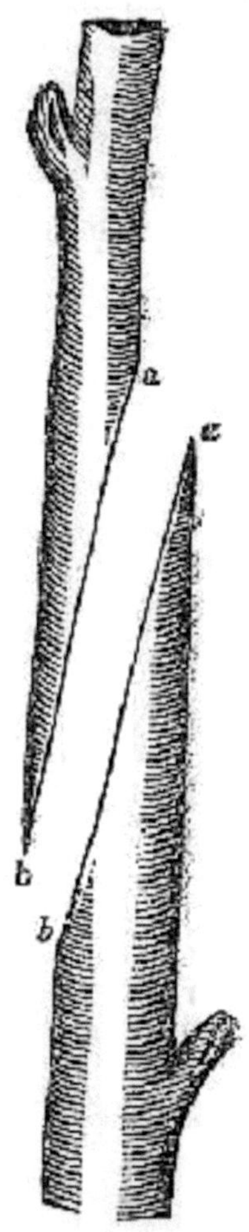

Fig. 41.

The time for ingrafting, is from May to October. The cuttings must be taken from horizontal shoots, between Christmas and March, and kept in a damp cellar. In performing the operation, cut off, in a sloping direction, (as seen in Fig. 41,) the tree or limb to be grafted. Then cut off, in a corresponding slant, the slip to be grafted on. Then put them together, so that the inner[Pg 345] bark of each shall match, exactly, on one side, and tie them firmly together, with woollen yarn. It is not essential that both be of equal size; if the bark of each meet together exactly on *one* side, it answers the purpose. But the two must not differ much, in size. The slope should be an inch and a half, or more, in length. After they are tied together, the place should be covered with a salve or composition of beeswax and rosin. A mixture of clay and cowdung will answer the same purpose. This last must be tied on with a cloth. Grafting is more convenient than budding, as grafts can be sent from a great distance; whereas buds must be taken in July or August, from a shoot of the present year's growth, and cannot be sent to any great distance.

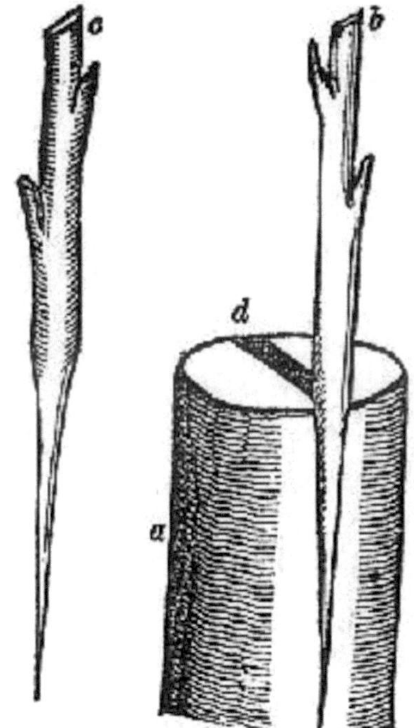

Fig. 42.

This engraving, (Fig. 42,) exhibits the mode called stock-grafting; *a*, being the limb of a large tree which is sawed off and split, and is to be held open by a small wedge, till the grafts are put in. A graft, inserted in the limb, is shown at *b*, and at *c*, is one not inserted, but designed to be put in at *d*, as two grafts can be put into a large stock. In inserting the graft, be careful to make the edge of the inner bark of the graft meet exactly the edge of the inner bark of the stock; for on this, success depends. After the grafts are put in, the wedge must be withdrawn, and the whole of the stock[Pg 346] be covered with the thick salve or composition before mentioned, reaching from where the grafts are inserted, to the bottom of the slit. Be careful not to knock or move the grafts, after they are put in.

Pruning.

The following rules for pruning, are from a distinguished horticulturist. Prune off all dead wood, and all the little twigs on the main limbs. Retrench branches, so as to give light and ventilation to the interior of the tree. Select the straight and perpendicular shoots, which give little or no fruit, while those which are most nearly horizontal, and somewhat curving, give fruit abundantly, and of good quality. Superfluous and ill-placed buds may be rubbed off, at any time; and no buds, pushing out after Midsummer, should be spared. In choosing between shoots to be retained, preserve the lowest placed; and, on lateral shoots, those which are nearest the origin. When branches cross each other, so as to rub, remove one or the other. Remove all suckers from the roots of trees or shrubs. Prune after the sap is in full circulation, (except in the case of grapes,) as the wounds then heal best. Some think it best to prune before the sap begins to run. Pruning-shears, and a pruning-pole, with a chisel at the end, can be procured of those who deal in agricultural utensils.

Thinning.

As it is the office of the leaves to absorb nourishment from the atmosphere, they should never be removed, except to mature the wood or fruit. In doing this, remove such leaves as shade the fruit, as soon as it is ready to ripen. To do it earlier, impairs the growth. Do it gradually, at two different times. Thinning the fruit is important, as tending to increase its size and flavor, and also to promote the longevity of the tree. If the fruit be thickly set, take off one half, at the time of setting. Revise in June, and then in July, taking off all that may be spared. One *very large* apple to[Pg 347] every square foot, is a rule that may be a sort of guide, in other cases. According to this, two hundred large apples would be allowed to a tree, whose extent is fifteen feet by twelve. If any person think this thinning excessive, let him try two similar trees, and thin one as directed, and leave the other unthinned. It will be found that the thinned tree will produce an equal weight, and fruit of much finer flavor.

CHAPTER XXXVI.
ON THE CULTIVATION OF FRUIT.

By a little attention to this matter, a lady, with the help of her children, can obtain a rich abundance of all kinds of fruit. The writer has resided in families, where little boys, of eight, ten, and twelve years old, amused themselves, under the direction of their mother, in planting walnuts, chestnuts, and hazelnuts, for future time; as well as in planting and inoculating young fruit-trees, of all descriptions. A mother, who will take pains to inspire a love for such pursuits, in her children, and who will aid and superintend them, will save them from many temptations; and, at a trifling expense, secure to them and herself a rich reward, in the choicest fruits. The information given in this work, on this subject, may be relied on, as sanctioned by the most experienced nursery-men.

The soil, for a nursery, should be rich, well dug, dressed with well-decayed manure, free from weeds, and protected from cold winds. Fruit seeds should be planted in the Autumn, an inch and a half or two inches deep, in ridges four or five feet apart, pressing the earth firmly over the seeds. While growing, they should be thinned out, leaving the best ones a foot and a half apart. The soil should be kept loose, soft, and free from weeds. They should be inoculated or [Pg 348]ingrafted, when of the size of a pipe stem; and in a year after this, may be transplanted to their permanent stand. Peach trees sometimes bear in two years from budding, and in four years from planting, if well kept.

In a year after transplanting, take pains to train the head aright. Straight, upright branches, produce *gourmands*, or twigs bearing only leaves. The side branches, which are angular or curved, yield the most fruit. For this reason, the limbs should be trained in curves, and perpendicular twigs should be cut off, if there be need of pruning. The last of June is the time for this. Grass should never be allowed to grow within four feet of a large tree, and the soil should be kept loose, to admit air to the roots. Trees in orchards should be twenty-five feet apart. The soil *under* the top soil, has much to do with the health of trees. If it be what is called *hard-pan*, the trees will deteriorate. Trees need to be manured, and to have the soil kept open and free from weeds.

Filberts can be raised in any part of this Country. *Figs* can be raised in the Middle States. For this purpose, in the Autumn, loosen the roots, on one side, and bend the tree down to the earth, on the other; then cover it with a mound of straw, earth, and boards; and early in the Spring raise it up, and cover the roots. *Currants* grow well in any but a wet soil. They are propagated by cuttings. The old wood should be thinned in the Fall, and manure be put on. They can be trained into small trees. *Gooseberries* are propagated by layers and cuttings. They are best, when kept from suckers and trained like trees. One third of the old wood should be removed every Autumn. *Raspberries* do best, when shaded during a part of the day. They are propagated by layers, slips, and suckers. There is one kind, which bears monthly. *Strawberries* require a light soil and vegetable manure. They should be transplanted in April or September, and be set eight inches apart, in rows nine inches asunder, and in beds which are two feet wide, with narrow alleys[Pg 349] between them. A part of these plants are *non-bearers.* These have large flowers, with showy stamens and high black anthers. The *bearers* have short stamens, a great number of pistils, and the flowers are every way less showy. In blossom-time, pull out all the non-bearers. Some think it best to leave one non-bearer to every twelve bearers; but others pull them all out. Many beds never produce any fruit, because all the plants in them are non-bearers. Weeds should be kept from the vines. When the vines are matted with young plants, the best way is to dig over the beds, in cross lines, so as to leave some of the plants standing in little squares, while the rest are turned under the soil. This should be done over a second time in the same year.

Grapes. To raise this fruit, manure the soil, and keep it soft, and free from weeds. A gravelly or sandy soil, and a south exposure, are best. Transplant the vines in the early Spring, or, better, in the Fall. Prune them, the first year, so as to have only two main branches, taking off all other shoots, as fast as they come. In November, cut off all of these two branches, except four eyes. The second year, in the Spring, loosen the earth around the roots, and allow only two branches to grow, and every month, take off all side shoots. When they are very strong, preserve only a part, and cut off the rest in the

Fall. In November, cut off all the two main stems, except eight eyes. After the second year no more pruning is needed, except to reduce the side shoots, for the purpose of increasing the fruit. All the pruning of grapes, (except nipping side shoots,) must be done when the sap is not running, or they will bleed to death. Train them on poles, or lattices, to expose them to the air and sun. Cover tender vines in the Autumn. Grapes are propagated by cuttings, layers, and seeds. For cuttings, select, in the Autumn, well-ripened wood, of the former year, and take five joints for each. Bury them, till April; then soak them, for some hours, and set them out, *aslant,* so that all the eyes but one shall be covered.

[Pg 350]

To Preserve Fruit.

Raspberries and Strawberries can be preserved, in perfect flavor, in the following manner. Take a pound of nice sifted sugar for each pound of fruit. Put them in alternate layers, of fruit and sugar, till the jar is entirely full, then cork it, and seal it air tight.

Currants and Gooseberries may be perfectly preserved thus. Gather them, when dry, selecting only the solid ones. Take off the stalks, and put them in dry junk-bottles. Set them, *uncorked,* in a kettle of water, and slowly raise it to boiling heat, in order to drive the air out of the bottles. Then take out the bottles, cork them, and seal them air tight. Keep them in a dry place, where they will not freeze. The success of this method depends on excluding air and water.

Apples, Grapes, and such like fruit can be preserved, by packing them, when dry and solid, in dry sand or sawdust, putting alternate layers of fruit and sawdust or sand. Some sawdust gives a bad flavor to the fruit.

Modes of Preserving Fruit Trees.

Heaps of ashes, or tanner's bark, around peach trees, prevent the attack of the worm. The *yellows,* is a disease of peach trees, which is spread by the pollen of the blossom. When a tree begins to turn yellow, take it away, with all its roots, before it blossoms again, or it will infect other trees. Planting tansy around the roots of fruit trees, is a sure protection against worms, as it prevents the moth from depositing her egg. Equal quantities of salt and saltpetre, put around the trunk of a peach tree, half a pound to a tree, improves the size and flavor of the fruit. Apply this about the first of April, and if any trees have worms already in them, put on half the quantity, in addition, in June. To young trees, just set out, apply one ounce, in April, and another in June, close to the stem. Sandy soil is best for peaches.

Apple trees are preserved from insects, by a wash of strong ley to the body and limbs, which, if old, should[Pg 351] be first scraped. Caterpillars should be removed, by cutting down their nests in a damp day. Boring a hole, in a tree infested with worms, and filling it with sulphur, will often drive them off immediately.

The *fire-blight*, or *brûlure*, in pear trees, can be stopped, by cutting off all the blighted branches. It is supposed, by some, to be owing to an excess of sap, which is remedied by diminishing the roots.

The *curculio*, which destroys plums, and other stone fruit, can be checked only by gathering up all the fruit that falls, (which contains their eggs,) and destroying it. The *canker-worm* can be checked, by applying a bandage around the body of the tree, and every evening smearing it with fresh tar.

CHAPTER XXXVII.
MISCELLANEOUS DIRECTIONS.

Every woman should know how to direct in regard to the proper care of domestic animals, as they often suffer from the negligence of domestics.

The following information, in reference to the care of a horse and cow, may be useful. A stable should not be very light nor very dark; its floor should be either plank or soil, as brick or stone pavements injure the feet. It should be well cleaned, every morning. A horse, kept in a stable, should be rubbed and brushed every day. A stable-horse needs as much daily exercise as trotting three miles will give him. Food or drink should never be given, when a horse is very warm with exercise, as it causes disease. A horse should be fed, three times a day. Hay, sheaf-oats, shorts, corn-meal, and bran, are the best food for horses. When a horse is travelling, order six quarts of oats in the morning, four at noon, and six at night, and direct that neither food nor water be given till he is cool.

[Pg 352]

Keep a horse's legs free from mud, or disease will often result from the neglect. A horse, much used, should be shod as often as once in two months. Fish-oil and strong perfumes, on the skin, keep flies from annoying a horse. Some horses are made fractious by having the check-rein so tight as to weary the muscles.

A cow should be watered three times a day, and fed with hay, potatoes, carrots, and boiled corn. Turnips and cabbages give a bad taste to the milk. Give a handful of salt to a cow, twice a week, and occasionally give the same quantity to a horse. Let them drink *pure* water. A well-fed cow gives double the milk that she will if not fed well. A cow should go unmilked, for two months before calving, and her milk should not be used till four days after. The calf must run with the cow for four days, and then be shut from her, except thrice a day, when it should take as much food as it wants, and then the cow should be milked clean.

Hens sit twenty days, and should be well fed and watered, during this time. The first food for chickens should be coarse dry meal.

Cold and damp weather is bad for all young fowls, and they should be well protected from it. Pepper-berries are good for fowls which have diseases caused by damp and cold weather.

In Winter, much fuel may be saved, and comfort secured, by stuffing cotton into all cracks about the windows and the surbases of rooms, and by listing the doors. Cover strips of wood with baize, and nail them tight against a door, on the casing.

The following are the causes of smoky chimneys. Short and broad flues, running up straight, as a narrow flue, with a bend in it, draws best. Large openings, at the top, draw the wind down, and should be remedied, by having the summits made tapering. A house higher than a chimney near it, sometimes makes the chimney smoke, and the evil should be remedied, by raising the chimney. Too large a throat to the fireplace, sometimes causes a chimney to smoke, and can be remedied, [Pg 353]by a false back, or by lowering the front, with sheet iron. Shallow fireplaces give out more heat, and draw as well, as deep ones.

House-cleaning should be done in dry warm weather. Several friends of the writer maintain, that cleaning paint, and windows, and floors, in *hard, cold* water, without any soap, using a flannel washcloth, is much better than using warm suds. It is worth trying. In cleaning in the common way, sponges are best for windows, and clean water only should be used. They should be first wiped with linen, and then with old silk. The outside of windows should be washed with a long brush, made for the purpose; and they should be rinsed, by throwing upon them water, containing a little saltpetre.

When inviting company, mention, in the note, the day of the month and week, and the hour for coming. Provide a place for ladies to dress their hair, with a glass, pins, and combs. A pitcher of cold water, and a tumbler, should be added. When the company is small, it is becoming a common method for the table to be set at one end of the room, the lady of the house to pour out tea, and the gentlemen of the party to wait on the ladies and themselves. When tea is sent round, always send a teapot of hot water to weaken it, and a slop-bowl, or else many persons will drink their tea much stronger than they wish.

Let it ever be remembered, that the burning of lights and the breath of guests, are constantly exhausting the air of its healthful principle; therefore avoid crowding many guests into one room. Do not tempt the palate by a great variety of unhealthful dainties. Have a warm room for departing guests, that they may not become chilled before they go out.

A parlor should be furnished with candle and fire screens, for those who have weak eyes; and if, at table, a person sits with the back near the fire, a screen should be hung on the back of the chair, as it is very injurious to the whole system to have the back heated.

[Pg 354]

Pretty baskets, for flowers or fruits, on centre tables, can be made thus. Knit, with coarse needles, all the various shades of green and brown, into a square piece. Press it with a hot iron, and then ravel it out. Buy a pretty shaped wicker basket, or make one of stiff millinet, or thin pasteboard, cut the worsted into bunches, and sew them on, to resemble moss. Then line the basket, and set a cup or dish of water in it, to hold flowers, or use it for a fruit-basket. Handsome fireboards are made, by nailing black foundation-muslin to a frame the size of the fireplace; and then cutting out flowers, from wall-paper, and pasting them on the muslin, according to the fancy.

India rubber, melted in lamp-oil, and brushed over common shoes, keeps water out, perfectly. Keep small whisk brooms, wherever gentlemen hang their clothes, both up stairs and down, and get them to use them if you can.

Boil new earthen in bran-water, putting the articles in, when cold. Do the same with porcelain kettles. Never leave wooden vessels out of doors, as they fall to pieces. In Winter, lift the handle of a pump, and cover it with blankets, to keep it from freezing.

Broken earthen and china, can often be mended, by tying it up, and boiling it in milk. *Diamond cement*, when genuine, is very effectual for the same purpose. Old putty can be softened by muriatic acid. Nail slats across nursery windows. Scatter ashes on slippery ice, at the door; or rather, remove it. Clarify impure water with powdered alum, a teaspoonful to a barrel.

[Pg 354a]

NOTE.

A volume, entitled the *American Housekeeper's Receipt Book*, prepared by the author of this work, under the supervision of several experienced housekeepers, is designed as a Supplement to this treatise on Domestic Economy. The following Preface and Analysis of the Contents will indicate its design more fully:

Preface (for the American Housekeeper's Receipt Book.)

The following objects are aimed at in this work:

First, to furnish an *original* collection of receipts, which shall embrace a great variety of simple and well-cooked dishes, designed for every-day comfort and enjoyment.

Second, to include in the collection only such receipts as have been tested by superior housekeepers, and warranted to be *the best*. It is not a book made up in *any* department by copying from other books, but entirely from the experience of the best practical housekeepers.

Third, to express every receipt in language which is short, simple, and perspicuous, and yet to give all directions so minutely as that the book can be kept in the kitchen, and be used by any domestic who can read, as a guide in *every one* of her employments in the kitchen.

Fourth, to furnish such directions in regard to small dinner-parties and evening company as will enable any young housekeeper to perform her part, on such occasions, with ease, comfort, and success.

Fifth, to present a good supply of the rich and elegant dishes demanded at such entertainments, and yet to set forth so large and tempting a variety of what is safe, healthful, and good, in connexion with such warnings and suggestions as it is hoped may avail to promote a more healthful fashion in regard both to entertainments and to daily table supplies. No book of this kind will sell without an adequate supply of the rich articles which custom requires, and in furnishing them, the writer has aimed to follow the example of

Providence, which scatters profusely both good and ill, and combines therewith the caution alike of experience, revelation, and conscience, "choose ye that which is good, that ye and your seed may live."

Sixth, in the work on Domestic Economy, together with this, to which it is a Supplement, the writer has attempted to secure, in a cheap and popular form, for American housekeepers, a work similar to an English work which she has examined, entitled the *Encyclopædia of Domestic Economy, by Thomas Webster and Mrs. Parkes*, containing over twelve hundred octavo pages of closely-printed matter, treating on every department of Domestic Economy; a work which will be found much more useful to English women, who have a plenty of money and well-trained servants, than to American housekeepers. It is believed that most in that work which would be of any practical use to American housekeepers, will be found in this work and the Domestic Economy.

Lastly, the writer has aimed to avoid the defects complained of by most housekeepers in regard to works of this description issued in this country, or sent from England, such as that, in some cases, the receipts are so rich as to be both expensive and unhealthful; in others, that they are so vaguely expressed as to be very imperfect guides; in [Pg 354b]others, that the processes are so elaborate and *fussing* as to make double the work that is needful; and in others, that the topics are so limited that some departments are entirely omitted, and all are incomplete.

In accomplishing these objects, the writer has received contributions of the pen, and verbal communications from some of the most judicious and practical housekeepers, in almost every section of this country, so that the work is fairly entitled to the name it bears of the *American* Housekeeper's Receipt Book.

The following embraces most of the topics contained in this work.

Suggestions to young housekeepers in regard to style, furniture, and domestic arrangements.

Suggestions in regard to different modes to be pursued both with foreign and American domestics.

On providing a proper supply of family stores, on the economical care and use of them, and on the furniture and arrangement of a store-closet.

On providing a proper supply of utensils to be used in cooking, with drawings to illustrate.

On the proper construction of ovens, and directions for heating and managing them.

Directions for securing good yeast and good bread.

Advice in regard to marketing, the purchase of wood, &c.

Receipts for breakfast dishes, biscuits, warm cakes, tea cakes, &c.

Receipts for puddings, cakes, pies, preserves, pickles, sauces, cat-sups, and also for cooking all the various kinds of meats, soups, and vegetables.

The above receipts are arranged so that the more healthful and simple ones are put in one portion, and the richer ones in another.

Healthful and favourite articles of food for young children.

Receipts for a variety of temperance drinks.

Directions for making tea, coffee, chocolate, and other warm drinks.

Directions for cutting up meats, and for salting down, corning, curing, and smoking.

Directions for making butter and cheese, as furnished by a practical and scientific manufacturer of the same, of Goshen, Conn., that land of rich butter and cheese.

A guide to a selection of a regular course of family dishes, which will embrace *a successive variety*, and unite convenience with good taste and comfortable living.

Receipts for articles for the sick, and drawings of conveniences for their comfort and relief.

Receipts for articles for evening parties and dinner parties, with drawings to show the proper manner of setting tables, and of supplying and arranging dishes, both on these, and on ordinary occasions.

An outline of arrangements for a family in moderate circum-stances, embracing the systematic details of work for each domestic, and the proper mode of doing it, as furnished by an accomplished housekeeper.

Remarks on the different nature of food and drinks, and their relation to the laws of health.

Suggestions to the domestics of a family, designed to promote a proper appreciation of the dignity and importance of their station, and a cheerful and faithful performance of their duties.

Miscellaneous suggestions and receipts.

[Pg 355]

A GLOSSARY
OF SUCH WORDS AND PHRASES AS MAY NOT EASILY BE UNDERSTOOD BY THE YOUNG READER.

[Many words, not contained in this Glossary, will be found explained in the body of the Work, in the places where they first occur. For these, see Index.]

Academy, the Boston, an association in Boston, established for the purpose of promoting the study and culture of the art of music.

Action brought by the Commonwealth, a prosecution conducted in the name of the public, or by the authority of the State.

Alcoholic, made of, or containing, alcohol, an inflammable liquid, which is the basis of ardent spirits.

Alkali, (plural *alkalies*,) a chemical substance, which has the property of combining with, and neutralizing the properties of, acids, producing salts by the combination. Alkalies change most of the vegetable blues and purples to green, red to purple, and yellow to brown. *Caustic alkali*, an alkali deprived of all impurities, being thereby rendered more caustic and violent in its operation. This term is usually applied to pure potash. *Fixed alkali*, an alkali that emits no characteristic smell, and cannot be volatilized or evaporated without great difficulty. Potash and soda are called the fixed alkalies. Soda is also called a *fossil*, or *mineral*, *alkali*, and potash, the *vegetable alkali*. *Volatile alkali*, an elastic, transparent, colorless, and consequently invisible gas, known by the name of ammonia, or ammoniacal gas. The odor of spirits of hartshorn is caused by this gas.

Anglo-American, English-American, relating to Americans descended from English ancestors.

Anne, Queen, a Queen of England, who reigned from A. D. 1702, to 1714. She was the daughter of James II., and succeeded to the throne on the death of William III. She died, August 1, 1714, in the fiftieth year of her age. She was not a woman of very great intellect; but was deservedly popular, throughout her reign, being a model of conjugal and maternal duty, and always intending to do good. She was honored with the title of 'Good Queen Anne', which showed the opinion entertained of her virtues by the people.

Anotta, Annotto, Arnotta, or *Rocou,* a soft, brownish-red substance, prepared from the reddish pulp surrounding the seeds of a tree, which grows in the West Indies, Guiana, and other parts of South America, called the *Bixa orellana.* It is used as a dye.

Anther, that part of the stamen of a flower which contains the pollen or farina, a sort of mealy powder or dust, which is necessary to the production of the flower.

Anthracite, one of the most valuable kinds of mineral coal, containing no bitumen. It is very abundant in the United States.

Aperient, opening.

[Pg 356]

Apple-corer, an instrument lately invented for the purpose of divesting apples of their cores.

Arabic, gum, see *Gum Arabic.*

Archæology, a discourse or treatise on antiquities.

Arnotto, see *Anotta.*

Arrow-root, a white powder, obtained from the fecula or starch of several species of tuberous plants in the East and West Indies, Bermuda, and other places. That from Bermuda is most highly esteemed. It is used as an article for the table, in the form of puddings; and also as a highly-nutritive, easily-digested, and agreeable, food, for invalids. It derives its name from having been originally used by the Indians, as a remedy for the poison of their arrows, by mashing and applying it to the wound.

Articulating process, the protuberance, or projecting part of a bone, by which it is so joined to another bone, as to enable the two to move upon each other.

Asceticism, the state of an ascetic, or hermit, who flies from society and lives in retirement, or who practises a greater degree of mortification and austerity than others do, or who inflicts extraordinary severities upon himself.

Astral lamp, a lamp, the principle of which was invented by Benjamin Thompson, (a native of Massachusetts, and afterwards Count Rumford,) in which the oil is contained in a large horizontal ring, having, at the centre, a burner, which communicates with the ring by tubes. The ring is placed a little below the level of the flame, and, from its large surface, affords a supply of oil for many hours.

Astute, shrewd.

Auld Robin Gray, a celebrated Scotch song, in which a young woman laments her having married an old rich man, whom she did not love, for the sake of providing for her poor parents.

Auricles, (from a Latin word, signifying the ear,) the name given to two appendages of the heart, from their fancied resemblance to the ear.

Baglivi, (George,) an eminent physician, who was born at Ragusa, in 1668, and was educated at Naples and Paris. Pope Clement XIV., on the ground of his great merit, appointed him, while a very young man, Professor of Anatomy and Surgery in the College of Sapienza, at Rome. He wrote several works, and did much to promote the cause of medical science. He died, A. D. 1706.

Bass, or bass wood, a large forest tree of America, sometimes called the lime-tree. The wood is white and soft, and the bark is sometimes used for bandages, as mentioned in page 343.

Beau Nash, see *Nash*.

Bell, Sir Charles, a celebrated surgeon, who was born in Edinburgh, in the year 1778. He commenced his career in London, in 1806, as a lecturer on Anatomy and Surgery. In 1830, he received the honors of knighthood, and in 1836 was appointed Professor of Surgery in the College of Edinburgh. He died near Worcester, in England, April 29, 1842. His writings are very numerous, and have been much celebrated. Among the most important of these, to general readers, are, his Illustrations of Paley's Natural Theology, (which work forms the second and third volumes of the larger series of 'The School Library,' issued by the Publishers of this volume,) and his treatise on 'The Hand, its Mechanism, and Vital Endowments, as evincing Design.'

Bergamot, a fruit, which was originally produced by ingrafting a branch of a citron or lemon tree, upon the stock of a peculiar kind of pear, called the bergamot pear.

Biased, cut diagonally from one corner to another of a square or rectangular piece of cloth. *Bias pieces*, triangular pieces cut as above mentioned.

Bituminous, containing *bitumen*, which is an inflammable mineral substance, resembling tar or pitch in its properties and uses. Among different bituminous substances, the names *naphtha* and *petroleum* have been given to those which are fluid; *maltha*, to that which has the consistence of pitch; and *asphaltum*, to that which is solid.

Blight, a disease in plants, by which they are blasted, or prevented from producing fruit.

Blond lace, lace made of silk.

Blood heat, the temperature which the blood is always found to maintain, or ninety-eight degrees of Fahrenheit's thermometer.

Blue vitriol, sulphate of copper. See *Sulphate*.

Blunts, needles of a short and thick shape, distinguished from *Sharps*, which are long and slender.

Bocking, a kind of thin carpeting, or coarse baize.

Boston Academy, see *Academy*.

Botany, (from a Greek word, signifying an herb,) a knowledge of plants; the science which treats of plants.

Brazil wood, the central part, or heart, of a large tree which grows in Brazil, called the *Cæsalpinia echinata*. It produces very lively and beautiful red tints, but they are not permanent.

Bronze, a metallic composition, consisting of copper and tin.

Brûlure, a French term, denoting a burning or scalding; a blasting of plants.

Brussels, (carpet,) a kind of carpeting, so called from the city of Brussels, in Europe. Its basis is composed of a warp and woof of strong linen threads, with the warp of which are intermixed about five times the quantity of woollen threads, of different colors.

Bulb, a root with a round body, like the onion, turnip, or hyacinth. *Bulbous*, having a bulb.

Byron, (George Gordon,) *Lord*, a celebrated Poet, who was born in London, January 22, 1788, and died in Missolonghi, in Greece, April 18, 1824.

Calisthenics, see page 56, note.

Camwood, a dyewood, procured from a leguminous (or pod-bearing) tree, growing on the Western Coast of Africa, and called *Baphia nitida*.

Cankerworm, a worm which is very destructive to trees and plants. It springs from an egg deposited by a miller that issues from the ground, and in some years destroys the leaves and fruit of apple and other trees.

Carbon, a simple inflammable body, forming the principal part of wood and coal, and the whole of the diamond.

Carbonic acid, a compound gas, consisting of carbon and oxygen. It has lately been obtained in a solid form.

Carmine, a crimson color, the most beautiful of all the reds. It is prepared from a decoction of the powdered cochineal insect, to which alum and other substances are added.

[Pg 358]

Caster, a small phial or vessel for the table, in which to put vinegar, mustard, pepper, &c.

Chancellor of the Exchequer, the highest judge of the law; the principal financial minister of a government, and the one who manages its revenue.

Chateau, a castle, a mansion.

Chemistry, the science which treats of the elementary constituents of bodies.

Chinese belle, deformities of. In China, it is the fashion to compress the feet of female infants, to prevent their growth; in consequence of which, the feet of all the females of China are distorted, and so small, that the individuals cannot walk with ease.

Chloride, a compound of chlorine and some other substance. *Chlorine* is a simple substance, formerly called oxymuriatic acid. In its pure state, it is a gas, of green color, (hence its name, from a Greek word, signifying green.) Like oxygen, it supports the combustion of some inflammable substances. *Chloride of lime* is a compound of chlorine and lime.

Cholera infantum, a bowel complaint, to which infants are subject.

Chyle, a white juice, formed from the chyme, and consisting of the finer and more nutritious parts of the food. It is afterwards converted into blood.

Chyme, the result of the first process which food undergoes in the stomach, previously to its being converted into chyle.

Cicuta, the common American Hemlock, an annual plant of four or five feet in height, and found commonly along walls and fences, and about old ruins and buildings. It is a virulent poison, as well as one of the most important and valuable medicinal vegetables. It is a very different plant from the Hemlock tree, or *Pinus Canadensis*.

Clarke, (Sir Charles Mansfield,) *Dr.*, a distinguished English physician and surgeon, who was born in London, May 28, 1782. He was appointed Physician to Queen Adelaide, wife of King William IV., in 1830, and in 1831, he was created a baronet. He is the author of several valuable medical works.

Cobalt, a brittle metal, of a reddish-gray color and weak metallic lustre, used in coloring glass. It is not easily melted nor oxidized in the air.

Cochineal, a color procured from the cochineal insect, (or *Coccus cacti*,) which feeds upon the leaves of several species of the plant called cactus, and which is supposed to derive its coloring matter from its food. Its natural color is crimson; but by the addition of a preparation of potash, it yields a rich scarlet dye.

Cologne water, a fragrant perfume, which derives its name from having been originally made in the city of Cologne, which is situated on the River Rhine, in Germany. The best kind is still procured from that city.

Comparative anatomy, the science which has for its object a comparison of the anatomy, structure, and functions, of the various organs of animals, plants, &c., with those of the human body.

Confection, a sweetmeat; a preparation of fruit with sugar; also a preparation of medicine with honey, sirup, or similar saccharine substance, for the purpose of disguising the unpleasant taste of the medicine.

Cooper, Sir Astley Paston, a celebrated English surgeon, who was born at Brooke, in Norfolk county, England, August 23, 1768, and [Pg 359]commenced the practice of Surgery in London, in 1792. He was appointed Surgeon to King George IV., in 1827, was created a baronet in 1821, and died February 12, 1841. He was the author of many valuable works.

Copal, a hard, shining, transparent resin, of a light citron color, brought, originally, from Spanish America, and now almost wholly from the East Indies. It is principally employed in the preparation of *copal varnish*.

Copper, sulphate of, see *Sulphate of copper.*

Copperas, (sulphate of iron, or green vitriol,) a bright green mineral substance, formed by the decomposition of a peculiar ore of iron, called pyrites, which is a sulphuret of iron. It is first in the form of a greenish-white powder, or crust, which is dissolved in water, and beautiful green crystals of copperas are obtained by evaporation. It is principally used in dyeing, and in making black ink. Its solution, mixed with a decoction of oak bark, produces a black color.

Coronary, relating to a crown or garland. In anatomy, it is applied to arteries which encompass the heart, in the manner, as it is fancied, of a garland.

Corrosive sublimate, a poisonous substance, composed of chlorine and quicksilver.

Cosmetics, preparations which some people foolishly think will preserve and beautify the skin.

Cream of tartar, see *Tartar.*

Crimping-iron, an instrument for crimping or curling ruffles, &c.

Curculio, a weevil or worm, which affects the fruit of the plum tree, and sometimes that of the apple tree, causing the unripe fruit to fall to the ground.

Curvature of the spine, see pages 80, 81.

Cuvier, Baron, the most eminent naturalist of the present age, was born, A. D. 1769, and died, A. D. 1832. He was Professor of Natural History in the College of France, and held various important posts under the French Government, at different times. His works on Natural History are of the greatest value.

Cynosure, the star near the North Pole, by which sailors steer. It is used, in a figurative sense, as synonymous with *pole-star*, or *guide*.

De Tocqueville, see *Tocqueville*.

Diamond cement, a cement sold in the shops, and used for mending broken glass, and similar articles.

Drab, a thick woollen cloth, of a light brown or dun color. The name is sometimes used for the color itself.

Dredging-box, a box with holes in the top, used to sift or scatter flour on meat, when roasting.

Drill, (in husbandry,) to sow grain in rows, drills, or channels; the row of grain so sowed.

Duchess of Orleans, see *Orleans*.

The *East*, and the *Eastern States*, those of the United States situated in the north-east part of the Country, including Maine, New Hampshire, Massachusetts, Rhode Island, Connecticut, and Vermont.

Electuary, a mixture, consisting of medicinal substances, especially dry powders, combined with honey or sirup, in order to render them less unpleasant to the taste, and more convenient for internal use.

[Pg 360]

Elevation, (of a house,) a plan, representing the upright view of a house, as a ground-plan shows its appearance on the ground.

Euclid, a celebrated mathematician, who was born in Alexandria, in Egypt, about two hundred and eighty years before Christ. He distinguished himself by his writings on music and geometry. The most celebrated of his works, is his 'Elements of Geometry,' which is in use at the present day. He established a school at Alexandria, which became so famous, that, from his time to the conquest of Alexandria by the Saracens, (A. D. 646,) no mathematician was found, who had not studied at Alexandria. Ptolemy, King of Egypt, was one of his pupils; and it was to a question of this King, whether there were not a shorter way of coming at Geometry, than by the study of his Elements, that Euclid made the celebrated answer, "There is no royal way, or path, to Geometry."

Equator, or *equinoctial line,* an imaginary line passing round the earth, from east to west, and directly under the sun, which always shines nearly perpendicularly down upon all countries situated near the equator.

Evolve, to throw off, to discharge.

Exchequer, a court in England, in which the Chancellor presides, and where the revenues of, and debts due to, the King are recovered. This court was originally established by King William, (called 'the Conqueror,') who died A. D. 1087; and its name is derived from a checkered cloth, (French *echiquier,* a chess-board, checker-work,) on the table.

Excretion, something discharged from the body, a separation of animal matters.

Excrementitious, consisting of matter excreted from the body; containing excrements.

Fahrenheit, (Gabriel Daniel,) a celebrated natural philosopher, who was born at Dantzic, A. D. 1686. He made great improvements in the thermometer; and his name is sometimes used for that instrument.

Farinaceous, mealy, tasting like meal.

To *Fell,* to turn down, on the wrong side, the raw edges of a seam, after it has been stitched, run, or sewed, and then to hem or sew it to the cloth.

Festivals, of the Jews, the three great annual. These were, the Feast of the Passover, that of Pentecost, and that of Tabernacles; on occasion of which, all the males of the Nation were required to visit the Temple at Jerusalem, in whatever part of the Country they might reside. See Exodus xxiii. 14, 17, xxxiv. 23, Leviticus xxiii. 4, Deuteronomy xvi. 16. The Passover was kept in commemoration of the deliverance of the Israelites from Egypt, and was so named, because, the night before their departure, the destroying angel, who slew all the first-born of the Egyptians, *passed over* the houses of the Israelites, without entering them. See Exodus xii. The Feast of Pentecost was so called, from a word meaning *the fiftieth*, because it was celebrated on the fiftieth day after the Passover, and was instituted in commemoration of the giving of the Law from Mount Sinai, on the fiftieth day from the departure out of Egypt. It is also called the Feast of Weeks, because it was kept seven weeks after the Passover. See Exodus xxxiv. 22, Leviticus xxiii. 15-21, Deuteronomy xvi. 9, 10. The Feast of Tabernacles, or Feast of Tents, was so called, because it was celebrated under tents or tabernacles of green boughs; and was designed to commemorate [Pg 361]their dwelling in tents, during their passage through the wilderness. At this Feast, they also returned thanks to God, for the fruits of the earth, after they had been gathered. See Exodus xxiii. 16, Leviticus xxiii. 34-44, Deuteronomy xvi. 13, and also St. John vii. 2.

Fire blight, a disease in the pear, and some other fruit trees, in which they appear burnt, as if by fire. It is supposed, by some, to be caused by an insect, others suppose it to be caused by an overabundance of sap.

Fluting-iron, an instrument for making flutes, channels, furrows, or hollows, in ruffles, &c.

Foundation muslin, a nice kind of buckram, stiff and white, used for the foundation or basis of bonnets, &c.

Free States, those States in which slavery is not allowed, as distinguished from Slave States, in which slavery does exist.

French chalk, a variety of the mineral called talc, unctuous to the touch, of a greenish color, glossy, soft, and easily scratched, and leaving a silvery line, when drawn on paper. It is used for marking on cloth, and extracting grease-spots.

Fuller's earth, a species of clay, remarkable for its property of absorbing oil; for which reason it is valuable for extracting grease from cloth, &c. It is used by fullers, in scouring and cleansing cloth, whence its name.

Fustic, the wood of a tree which grows in the West Indies, called *Morus tinctoria*. It affords a durable, but not very brilliant, yellow dye, and is also used in producing some greens and drab colors.

Gastric, (from the Greek γαστιρ, *gaster*, the belly,) belonging or relating to the belly, or stomach. *Gastric juice*, the fluid which dissolves the food in the stomach. It is limpid, like water, of a saltish taste, and without odor.

Geology, the science which treats of the earth, as composed of rocks and stones.

Gore, a triangular piece of cloth. *Goring*, cut in a triangular shape.

Gothic, a peculiar and strongly-marked style of architecture, sometimes called the ecclesiastical style, because it is most frequently used in cathedrals, churches, abbeys, and other religious edifices. Its principle seems to have originated in the imitation of groves and bowers, under which the ancients performed their sacred rites; its clustered pillars and pointed arches very well representing the trunks of trees and their interlocking branches.

Gourmand, or *Gormand*, a glutton, a greedy eater. In agriculture, it is applied to twigs which take up the sap, but bear only leaves.

Green vitriol, see *Copperas*.

Griddle, an iron pan, of a peculiarly broad and shallow construction, used for baking cakes.

Ground-plan, the map or plan of the lower floor of any building, in which the various apartments, windows, doors, fireplaces, and other things, are represented, like the rivers, towns, mountains, roads, &c., on a map.

Gum Arabic, a vegetable juice which exudes through the bark of the *Acacia, Mimosa nilotica*, and some other similar trees, growing in Arabia, Egypt, Senegal, and Central Africa. It is the purest of all gums.

Hardpan, the hard, unbroken layer of earth, below the mould or cultivated soil.

Hartshorn, (spirits of,) a volatile alkali, originally prepared from the [Pg 362]horns of the stag or hart, but now procured from various other substances. It is known by the name of ammonia, or spirits of ammonia.

Hemlock, see *Cicuta*.

Horticulturist, one skilled in horticulture, or the art of cultivating gardens; horticulture being to the garden, what agriculture is to the farm, the application of labor and science to a limited spot, for convenience, for profit, or for ornament,—though implying a higher state of cultivation, than is common in agriculture. It includes the cultivation of culinary vegetables and of fruits, and forcing or exotic gardening, as far as respects useful products.

Hoskin's gloves, gloves made by a person named Hoskin, whose manufacture was formerly much celebrated.

Hydrogen, a very light, inflammable gas, of which water is, in part, composed. It is used to inflate balloons.

Hypochondriasis, melancholy, dejection, a disorder of the imagination, in which the person supposes he is afflicted with various diseases.

Hysteria, or *hysterics*, a spasmodic, convulsive affection of the nerves, to which women are subject. It is somewhat similar to hypochondriasis in men.

Ingrain, a kind of carpeting, in which the threads are dyed in the grain, or raw material, before manufacture.

Ipecac, (an abbreviation of *ipecacuanha*,) an Indian medicinal plant, acting as an emetic.

Isinglass, a fine kind of gelatin, or glue, prepared from the swimming-bladders of fishes, used as a cement, and also as an ingredient in food and medicine. The name is sometimes applied to a transparent mineral substance called mica.

Kamtschadales, inhabitants of *Kamtschatka*, a large peninsula situated on the northeastern coast of Asia, having the North Pacific

Ocean on the east. It is remarkable for its extreme cold, which is heightened by a range of very lofty mountains, extending the whole length of the peninsula, several of which are volcanic. It is very deficient in vegetable productions, but produces a great variety of animals, from which the richest and most valuable furs are procured. The inhabitants are in general below the common height, but have broad shoulders and large heads. It is under the dominion of Russia.

Kink, a knotty twist in a thread or rope.

Lapland, a country at the extreme north part of Europe, where it is very cold. It contains lofty mountains, some of which are covered with perpetual snow and ice.

Latin, the language of the Latins, or inhabitants of Latium, the principal country of ancient Italy. After the building of Rome, that city became the capital of the whole country.

Leguminous, pod-bearing.

Lent, a fast of the Christian Church, (lasting forty days, from Ash Wednesday to Easter,) in commemoration of our Saviour's miraculous fast of forty days and forty nights, in the wilderness. The word Lent means spring; this fast always occurring at that season of the year.

Levite, one of the tribe of Levi, the son of Jacob, which tribe was set apart from the others, to minister in the services of the Tabernacle, and the Temple at Jerusalem. The Priests were taken from this tribe. See Numbers i. 47-53.

[Pg 363]

Ley, water which has percolated through ashes, earth, or other substances, dissolving and imbibing a part of their contents. It is generally spelled *lie*, or *lye*.

Linnæus, (Charles,) a native of Sweden, and the most celebrated naturalist of his age. He was born May 13, 1707, and died January 11, 1778. His life was devoted to the study of natural history. The science of botany, in particular, is greatly indebted to his labors. His 'Amœnitates Academicæ' (Academical Recreations) is a collection of the dissertations of his pupils, edited by himself; a work rich in matters relating to the history and habits of plants. He was the first who arranged Natural History into a regular system, which has been generally called by his name. His proper name was Linné.

Lobe, a division, a distinct part; generally applied to the two divisions of the lungs.

Log Cabin, a cabin or house built of logs, as is generally the case in newly-settled countries.

Loire, the largest river of France, being about five hundred and fifty miles in length. It rises in the mountains of Cevennes, and empties into the Atlantic Ocean, about forty miles below the city of Nantes. It divides France into two almost equal parts.

London Medical Society, a distinguished association, formed in 1773. It has published some valuable volumes of its Transactions. It has a library, of about 40,000 volumes, which is kept in a house presented to the Society, in 1788, by the celebrated Dr. Lettsom, who was one of its first members.

Louis XIV., a celebrated King of France and Navarre, who was born Sept. 5, 1638, and died Sept. 1, 1715. His mother having before had no children, though she had been married twenty-two years, his birth was considered as a particular favor from heaven, and he was called the 'Gift of God.' He is sometimes styled 'Louis the Great,' and his reign is celebrated as an era of magnificence and learning, and is notorious as a period of licentiousness. He left behind him monuments of unprecedented splendor and expense, consisting of palaces, gardens, and other like works.

Lumbar, (from the Latin *lumbus*, the loin,) relating or pertaining to the loins.

Lunacy, writ of, a judicial proceeding, to ascertain whether a person be a lunatic.

Mademoiselle, the French word for Miss, a young girl.

Magnesia, a light and white alkaline earth, which enters into the composition of many rocks, communicating to them a greasy or soapy feeling, and a striped texture, with sometimes a greenish color.

Malaria, (Italian, *mal'aria, bad air,*) a noxious vapor or exhalation; a state of the atmosphere or soil, or both, which, in certain regions, and in warm weather, produces fever, sometimes of great violence.

Mammon, riches, the Syrian god of riches. See St. Luke, xvi. 11, 13, St. Matthew, vi. 24.

Martineau, (Harriet,) a woman who has become somewhat celebrated by her book of travels in the United States, and by other works.

Mexico, a country situated southwest of the United States, and extending to the Pacific Ocean.

Miasms, such particles or atoms, as are supposed to arise from distempered, putrefying, or poisonous bodies.

[Pg 364]

Michilimackinac, or *Mackinac*, (now frequently corrupted into *Mackinaw*, which is the usual pronunciation of the name,) a military post in the State of Michigan, situated upon an island about nine miles in circuit, in the strait which connects Lakes Michigan and Huron. It is much resorted to by Indians and fur traders. The highest summit of the island is about three hundred feet above the lakes, and commands an extensive view of them.

Midsummer, with us, the time when the sun arrives at his greatest distance from the equator, or about the twenty-first of June, called, also, the summer solstice, (from the Latin *sol, the sun*, and *sto, to stop or stand still*,) because, when the sun reaches this point, he seems to stand still for some time, and then appears to retrace his steps. The days are then longer than at any other time.

Migrate, to remove from one place to another; to change residence.

Mildew, a disease of plants; a mould, spot, or stain, in paper, cloths, &c., caused by moisture.

Militate, to oppose, to operate against.

Millinet, a coarse kind of stiff muslin, formerly used for the foundation or basis of bonnets, &c.

Mineralogy, a science which treats of the inorganic natural substances found upon or in the earth, such as earths, salts, metals, &c., and which are called by the general name of minerals.

Minutiæ, the smallest particulars.

Monasticism, monastic life; religiously recluse life, in a monastery, or house of religious retirement.

Montagu, Lady Mary Wortley, one of the most celebrated among the female literary characters of England. She was daughter of Evelyn, Duke of Kingston, and was born about 1690, at Thoresby, in England. She displayed uncommon abilities, at a very early age, and was educated by the best masters in the English, Latin, Greek, and French, languages. She accompanied her husband (Edward Wortley Montagu) on an embassy to Constantinople, and her correspondence with her friends was published and much admired. She introduced the practice of inoculation for the smallpox into England, which proved of great benefit to millions. She died at the age of seventy-two, A. D. 1762.

Moral Philosophy, the science which treats of the motives and rules of human actions, and of the ends to which they ought to be directed.

Moreen, a kind of woollen stuff used for curtains, covers of cushions, bed hangings, &c.

Mucous, having the nature of *mucus*, a glutinous, sticky, thready, transparent fluid, of a salt savor, produced by different membranes of the body, and serving to protect the membranes and other internal parts against the action of the air, food, &c. The fluid of the mouth and nose is mucus.

Mucous membrane, that membrane which lines the mouth, nose, intestines, and other open cavities of the body.

Muriatic acid, an acid, composed of chlorine and hydrogen, called, also, hydrochloric acid, and spirit of salt.

Mush-stick, a stick to use in stirring *mush*, which is corn meal boiled in water.

Nankeen, or *Nankin*, a light cotton cloth, originally brought from Nankin, in China, whence its name.

[Pg 365]

Nash, (Richard,) commonly called *Beau Nash*, or King of Bath, a celebrated leader of the fashions in England. He was born at Swansea, in South Wales, October 8, 1674, and died in the city of Bath, (England,) February 3, 1761.

Natural History, the history of animals, plants, and minerals.

Natural Philosophy, the science which treats of the powers of Nature, the properties of natural bodies, and their action one upon another. It is sometimes called *physics*.

New-milch cow, a cow which has recently calved.

Newton, (Sir Isaac,) an eminent English philosopher and mathematician, who was born on Christmas day, 1642, and died March 20, 1727. He was much distinguished for his very important discoveries in Optics and other branches of Natural Philosophy. See the first volume of 'Pursuit of Knowledge under Difficulties,' forming the fourteenth volume of 'The School Library,' Larger Series.

Non-bearers, plants which bear no flowers nor fruit.

Northern States, those of the United States situated in the Northern and Eastern part of the Country.

Ordinary, see *Physician in Ordinary*.

Oil of Vitriol, (sulphuric acid, or vitriolic acid,) an acid composed of oxygen and sulphur.

Orleans, (Elizabeth Charlotte de Bavière,) *Duchess of*, second wife of Philippe, the brother of Louis XIV., was born at Heidelberg, May 26, 1652, and died at the palace of St. Cloud, in Paris, December 8, 1722. She was author of several works; among which were, Memoirs, and Anecdotes, of the Court of Louis XIV.

Ottoman, a kind of hassock, or thick mat, for kneeling upon; so called, from being used by the Ottomans or Turks.

Oxalic acid, a vegetable acid, which exists in sorrel.

Oxide, a compound (which is not acid) of a substance with oxygen; for example, oxide of iron, or rust of metals.

Oxidize, to combine oxygen with a body without producing acidity.

Oxygen, vital air, a simple and very important substance, which exists in the atmosphere, and supports the breathing of animals and the burning of combustibles. It was called oxygen, from two Greek words, signifying to produce acid, from its power of giving acidity to many compounds in which it predominates.

Oxygenized, combined with oxygen.

Pancreas, a gland within the abdomen, just below and behind the stomach, and providing a fluid to assist digestion. In animals, it is called the sweet-bread. *Pancreatic*, belonging to the pancreas.

Parterre, a level division of ground, a flower garden.

Pearlash, the common name for impure carbonate of potash, which, in a purer form, is called *Sal æratus*.

Peristaltic, worm-like.

Philosophy, see *Intellectual, Moral,* and *Natural.*

Physician in Ordinary to the Queen, the Physician who attends the Queen in ordinary cases of illness.

Pistil, that part of a flower, generally in the centre, composed of the germ, style, and stigma, which receives the pollen or fertilizing dust of the stamens.

Pitt, William, a celebrated English statesman, son of the Earl of Chatham. He was born, May 28, 1759, and at the age of twenty-three, was made Chancellor of the Exchequer, and soon afterward, Prime Minister. He died, January 23, 1806.

[Pg 366]

Political Economy, the science which treats of the general causes affecting the production, distribution, and consumption, of articles of exchangeable value, in reference to their effects upon national wealth and welfare.

Pollen, the fertilizing dust of flowers, produced by the stamens, and falling upon the pistils, in order to render a flower capable of producing fruit or seed.

Potter's clay, the clay used in making articles of pottery.

Prairie, a French word, signifying *meadow*. In the United States, it is applied to the remarkable natural meadows, or plains, which are found in the Western States. In some of these vast and nearly level plains, the traveller may wander for days, without meeting with wood or water, and see no object rising above the plane of the horizon. They are very fertile.

Prime Minister, the person appointed by the ruler of a nation to have the chief direction and management of the public affairs.

Process, a protuberance, or projecting part of a bone.

Pulmonary, belonging to, or affecting, the lungs. *Pulmonary artery*, an artery which passes through the lungs, being divided into several branches, which form a beautiful network over the air-vessels, and finally empty themselves into the left auricle of the heart.

Puritans, a sect, which professed to follow the pure word of God, in opposition to traditions, human constitutions, and other authorities. In the reign of Queen Elizabeth, part of the Protestants were desirous of introducing a simpler, and, as they considered it, a *purer*, form of church government and worship, than that established by law; from which circumstance, they were called *Puritans*. In process of time, this party increased in numbers, and openly broke off from the Church, laying aside the English liturgy, and adopting a service-book published at Geneva, by the disciples of Calvin. They were treated with great rigor by the Government, and many of them left the kingdom and settled in Holland. Finding themselves not so eligibly situated in that Country, as they had expected to be, a portion of them embarked for America, and were the first settlers of New England.

Quixotic, absurd, romantic, ridiculous; from *Don Quixote*, the hero of a celebrated fictitious work, written by Cervantes, a distinguished Spanish writer, and intended to reform the tastes and opinions of his countrymen.

Reeking, smoking, emitting vapor.

Residuum, the remainder, or part which remains.

Routine, a round, or course of engagements, business, pleasure, &c.

To *Run* a seam, to lay the two edges of a seam together, and pass the threaded needle out and in, with small stitches, a few threads below the edge, and on a line with it.

To *Run* a stocking, to pass a thread of yarn, with a needle, straight along each row of the stocking, as far as is desired, taking up one loop and missing two or three, until the row is completed, so as to double the thickness at the part which is run.

Sabbatical year, every seventh year, among the Jews, which was a year of rest for the land, when it was to be left without culture. In this year, all debts were to be remitted, and slaves set at liberty. See Exodus xxi. 2, xxiii. 10, Leviticus xxv. 2, 3, &c., Deuteronomy xv. 12, and other similar passages.

Sal æratus, see *Pearlash.*

[Pg 367]

Sal ammoniac, a salt, called also muriate of ammonia, which derives its name from a district in Libya, Egypt, where there was a temple of Jupiter Ammon, and where this salt was found.

Scotch Highlanders, inhabitants of the Highlands of Scotland.

Selvedge, the edge of cloth, a border. Improperly written *selvage.*

Service-book, a book prescribing the order of public services in a church or congregation.

Sharps, see *Blunts.*

Shorts, the coarser part of wheat bran.

Shrubbery, a plantation of shrubs.

Siberia, a large country in the extreme northern part of Asia, having the Frozen Ocean on the north, and the Pacific Ocean on the east, and forming a part of the Russian empire. The northern part is extremely cold, almost uncultivated, and contains but few inhabitants. It furnishes fine skins, and some of the most valuable furs in the world. It also contains rich mines of iron and copper, and several kinds of precious stones.

Sinclair, Sir John, of whom it was said, "There is no greater name in the annals of agriculture, than his," was born in Caithness, Scotland, May 10, 1754, and became a member of the British Parliament in 1780. He was strongly opposed to the measures of the British Government towards America, which produced the American Revolution. He was author of many valuable publications, on various subjects. He died December 21, 1835.

Sirloin, the loin of beef. The appellation 'Sir' is the title of a knight, or baronet; and has been added to the word 'loin,' when applied to beef, because a King of England, in a freak of good humor, once conferred the honor of knighthood upon a loin of beef.

Slack, to loosen, to relax, to deprive of cohesion.

Soda, an alkali, usually obtained from the ashes of marine plants.

To *Spade*, to throw out earth with a spade.

Spermaceti, an oily substance, found in the head of a species of whale, called the spermaceti whale.

Spindling, see page 124.

Spinous process, a process or bony protuberance, resembling a spine or thorn, whence it derives its name.

Spool, a piece of cane or reed, or a hollow cylinder of wood, with a ridge at each end, used to wind yarn and thread upon.

Stamen, (plural *stamens* and *stamina*,) in *weaving*, the warp, the thread, any thing made of threads. In *botany*, that part of a flower, on which the artificial classification is founded, consisting of the filament or stalk, and the anther, which contains the pollen, or fructifying powder.

Stigma, (plural *stigmas* and *stigmata*,) the summit or top of the pistil of a flower.

Style, or *Stile*, the part of the pistil between the germ and the stigma.

Sub-carbonate, an imperfect carbonate.

Sulphates, Sulphats, Sulphites, salts formed by the combination of some base with sulphuric acid, as *Sulphate of copper*, (blue vitriol, or blue stone,) a combination of sulphuric acid with copper. *Sulphate of iron,* copperas, or green vitriol. *Sulphate of lime,* gypsum, or plaster of Paris. *Sulphate of magnesia,* Epsom salts. *Sulphate of potash,* a chemical salt, composed of sulphuric acid and potash. *Sulphate of soda,* Glauber's salts. *Sulphate of zinc,* white vitriol.

[Pg 368]

Sulphuret, a combination of an alkaline earth or metal with sulphur as, *Sulphuret of iron,* a combination of iron and sulphur.

Sulphuric acid, oil of vitriol, vitriolic acid.

Suture, a sewing; the uniting of parts by stitching; the seam or joint which unites the flat bones of the skull, which are notched like the teeth of a saw, and the notches, being united together, present the appearance of a seam.

Tartar, a substance, deposited on the inside of wine casks, consisting chiefly of tartaric acid and potash. *Cream of tartar,* the crude tartar separated from all its impurities, by being dissolved in water and then crystallized, when it becomes a perfectly white powder.

Tartaric acid, a vegetable acid which exists in the grape.

Technology, a description of the arts, considered generally, in their theory and practice, as connected with moral, political, and physical science.

Three great Jewish yearly festivals, see *Festivals.*

Three-ply, or triple ingrain, a kind of carpeting, in which the threads are woven in such a manner as to make three thicknesses of the cloth.

Tic douloureux, a painful affection of the nerves, mostly those of the face.

Tocqueville, (Alexis de,) a celebrated living statesman and writer of France, and author of volumes on the Political Condition, and the Penitentiaries, of the United States, and other works.

Trachea, the windpipe, so named (from a Greek word signifying *rough*) from the roughness, or inequalities, of the cartilages of which it is formed.

Truckle-bed, or *trundle-bed,* a bed that runs on wheels.

Tuber, a solid, fleshy, roundish root, like the potato. *Tuberous,* thick and fleshy; composed of, or having, tubers.

Tucks, (improperly tacks,) folds in garments.

Turmeric, the root of a plant called *Curcuma longa,* a native of the East Indies, used as a yellow dye.

Twaddle, idle, foolish talk, or conversation.

Unbolted, unsifted.

Unslacked, not loosened, or deprived of cohesion. Lime, when it has been slacked, crumbles to powder, from being deprived of cohesion.

Valance, the drapery or fringe hanging round the cover of a bed, couch, or other similar article.

Vascular, relating to, or full of, vessels.

Venetian, a kind of carpeting, composed of a striped woollen warp on a thick woof of linen thread.

Verisimilitude, probability, resemblance to truth.

Verbatim, word for word.

Vice versa, the side being changed, or the question reversed, or the terms being exchanged.

Viscera, (plural of *viscus,*) organs contained in the abdomen and in the chest.

Vitriol, a compound mineral salt, of a very caustic taste. *Blue vitriol,* sulphate of copper. *Green vitriol,* see *Copperas. Oil of vitriol,* sulphuric acid. *White vitriol,* sulphate of zinc.

Waffle-iron, an iron utensil for the purpose of baking waffles, which are thin and soft cakes indented by the iron in which they are baked.

Washleather, a soft, pliable leather, dressed with oil, and in such a way, that it may be washed, without shrinking. It is used for various articles of dress, as under-shirts, drawers, &c., and also for rubbing [Pg 369]silver, and other articles having a high polish. The article known, in commerce, as chamois, or shammy, leather, is also called wash-leather.

Welting cord, a cord sewed into the welt or border of a garment.

The *West*, or *Western World*. When used in Europe, or in distinction from the Eastern World, it means America. When used in this Country, the West refers to the Western States of the Union. *Western Wilds*, the wild, thinly-settled lands of the Western States.

White vitriol, see *Zinc*.

Wilton carpet, a kind of carpets, made in England, and so called from the place which is the chief seat of their manufacture. They are woollen velvets, with variegated colors.

Writ of lunacy, see *Lunacy*.

Xantippe, the wife of Socrates, noted for her violent temper and scolding propensities. The name is frequently applied to a shrew, or peevish, turbulent, scolding woman.

Zinc, a blueish-white metal, which is used as a constituent of brass, and some other alloys. *Sulphate of zinc*, or *White vitriol*, a combination of zinc with sulphuric acid.

[Pg 371]